The Unifying Theory

By Forester de Santos

About The Unifying Theory

The Unifying Theory is about true knowledge through search and research of the physical world and not through abstract ideas or imaginary worlds or even invented science.

No PhD required, not even a doctorate or a master's degree to really enjoy these important and beneficial discoveries.

The Unifying Theory is for every conscious being no matter how great or how tall his consciousness or his mental abilities because true knowledge is for every conscious being no matter what his conscious or mental abilities.

True knowledge or confirmation is needed for birth or beginning or even creation and true acknowledgement or reconfirmation is needed for rebirth or for continuation or to refresh creation.

If there is no acknowledgement thus time stops and the universe ends with an implosion, a physical implosion and not an abstract one.

Tags: unifying, research, consciousness, universe, creation, physics, atom, black hole, implosion, matter, brain, big bang!

About this Author Beloved

All writers must come sooner or later to the things that they want to truly write about and most come to write where the easy money is and most writers make a very big killing in becoming very rich and very famous by writing fiction and good for them!

But the great question is how much fantasy for the human race since the human race has being living in a fantasy since the race entered into consciousness or began to think and invent with words?

Well, that is a great or tall question that Forester de Santos as a writer has truly asked!

And so he has truly chosen to walk or to actually write on the road less taken or less written about!

And so he began to research and write about immortality and as the same goes, one becomes what one thinks or writes or even reads about the most!

Prologue

Truly glad and truly joyful is that man that has true knowledge of how existence truly functions because that man also knows how that man himself truly functions.

And he also will do as existence her very self to continue on existing or living on but existing or living on as if eternally new and in complete abundance, all five portions of her…

Thus, Know the real function of the existence and you will function much better or with a greater or taller identity, because to come to know was because of really doing and thus really entering into understanding, which really is to enter into the right side of existence as the right side of existence and thus one becoming as the very beginning because of one renewing or because of one refreshing existence because of one knowing or one understanding...

Acknowledgement

I would like very much to give thanks to my youngest son for suggesting to me to write this short work about the unifying theory.

Thanks Freddie!

Dedication

The Unifying Theory is dedicated with all my heart and with my entire conscious mind to the good reader and to those who gladly and joyfully fight or gladly and joyfully struggle to understand or to find the truth and enter in the truth and thus become the truth herself.

Table of Contents

Introduction

Existence

(+) (0) (-)

Existence really exists because she cannot stop from existing. Existence has no choice but to only or to simply but exists!

Existence really is composed of three portions of lack but which can be seen and existence also is composed of one portion which really is but cannot be seen as of yet and which makes for three or even makes for four or for more portions...

Existence also is composed of opposites which really are more opposites of themselves than to the opposites to which they are opposites of.

And this opposition or contrast which makes possible for existence to attract her very self and it appears that existence moves or is in motion because of the very attraction of her very self.

And existence is known or is seen more through the little which is seen of her and through the more which is seen of her, less of her is known or less is seen...

And even though existence exists and she attracts herself and she also is eternal knowledge and also she gives knowledge to the vacuum of space or to the universe and thus creating the beginning or the times, existence does not know that she exists even though she also renews herself so that she can continue existing and existing as if always new and as if nothing has happened and will never happen...

But existence renews herself or is reborn when the vacuum of space becomes empty again or when the knowledge given to the universe is not taken as acknowledgement and she once again gives or once again throws more knowledge into the vacuum of space or into the space of the universe, thus making as if a new beginning and everything which was before as if it never had being or as if it never were...

But that ability which existence has of renewing herself and of continuing for all eternity without knowing end or ending thus we conscious beings also possess.

But we are not obligated to live eternity as neither we are obligated to death or to the end or the ending.

Death or the end or the ending comes because it is not done to continue but to continue as if new and to continue forever or to continue for all the times...

Those that do not want eternity thus they just wait for death and death will do for them so that there is no eternity for them...

But those of us that want to continue with life without seeing or without knowing death thus we must do to be reborn and that rebirth or being born again truly is done with knowledge or with acknowledgement.

Thus, we must give knowledge or give acknowledgement to that grandiose part of existence which is but which cannot be seen as of yet but even so it is what gives or it is what grants knowledge not only to the vacuum of space or to the universe but also it is what gives or what grants knowledge to every conscious being that requests it…

And just as one presents oneself to that grandiose part of existence which gives or which grants or which even lends knowledge, thus that grandiose part of existence will be to one.

In other words, that grandiose or glorious part is the Creating part of existence or the renovating part or the part which gives identity to all existence or which reacts with all the opposites of existence…

Thus, he that gives or that grants or that lends to the Creating part or to the Renovating part of existence the knowledge or the acknowledgement of God or of Creator or of Renovator thus he also will have the knowledge or the acknowledgement of God or of Renovator and God will present to him or come to him or will allow him to draw near with that very same knowledge or acknowledgement…

Existence gives or grants or lends to the vacuum of space or to the universe knowledge in the form of matter or stars or light.

And if that knowledge in the vacuum of space or in the universe is converted into acknowledgement thus the universe will continue as if forever new without ever knowing end or ending nor knowing or remembering beginning…

But if that knowledge given or granted or lend to the vacuum of space or to the universe does not renew or is not converted or is not transformed into acknowledgement, thus that knowledge in the form of matter or stars or light thus will lose her energy and there will only remain and will be space dust no matter how large the piece of space dust and there will only be darkness or there will only be emptiness or vacuum...

But that does not remain like that because now the dust which remained in the vacuum of space or in the vacuum of the universe must be taken out to give or to grant or to lend new knowledge to the vacuum of space or to the universe and that new knowledge will make a new beginning, in where there will not be any memory that there ever was a beginning before...

The manner in which existence takes the dust from the vacuum of space or from the universe is with black holes which truly are black vacuum cleaners.

The black vacuums or black holes also suck up any other matter or star which has remained in the vacuum of space or in the space of the universe...

Once there no longer is matter or dust in space thus the black holes also will turn of or will stop from functioning because of lack of energy and they will disperse or they will disintegrate in the vacuum of space.

This emptiness in space, which now has become as if a vacuum or new space because of becoming as before the beginning thus will attract new knowledge in form of matter or stars or light...

And if in that new beginning the same happens which happened in the first, in where there was no renovation or there was no acknowledgement or no rebirth to continue, thus also will have its end or ending even though it may take billions of years…

But that does not have to be as the above because as long as there are conscious beings in the universe, the universe has the very grandiose opportunity of renovating or of rebirth or of receiving acknowledgement so that the universe because of the conscious beings the universe will continue without ever knowing end or ending…

Existence without knowing it renews every time she lends knowledge in the form of matter or in the form of stars or of light to the vacuum of space or to the universe and the universe cannot continue for lack of acknowledgement or for lack of matter or for lack of renovated energy and thus the universe comes to its end or to its ending and thus making space for another beginning which will become as if the first beginning and also as if it never had an end or never an ending before…

But if the conscious beings are reborn or revive or take new life or receive acknowledgement through the very same knowledge or acknowledgement which they give or grant or lend, the universe will never ever have end or ending because the universe will continue as if forever new and as if it never had any beginning…

Now then, every conscious being truly has the very grandiose opportunity of being reborn or of reviving or of taking new life or of receiving acknowledgement or rename to be able to continue eternally with life and in complete harmony and in complete abundance…

But if the conscious being does not desire that very grandiose opportunity of living eternally and living in complete harmony and in complete abundance thus that conscious being only has to wait to die and that will he his end or his ending and nothing will become of that conscious being because eternity or immortality is not obligated or is not imposed, even a rock will stop from being or from existing…

Thus, one needs to be alive and conscious to be able to have or can receive immortality in the form of salvation and with her continue alive renewing and renewing also everything else as savior…

Just as knowledge in the form of matter filled with energy or in the form of stars or in the form of light enters the vacuum of space or enters into the universe, thus that same way also thought or knowledge or illumination enters the conscious mind.

That thought or that knowledge or that illumination can take the conscious being to a great state or from one state to another state or to a greater identity or from one identity to another identity even though that conscious being really continues with his physical form but every time that that conscious being enters into a greater state of knowledge or into a greater or new identity because of his knowledge, thus the physical form of that conscious being also is refreshed or is seen as if a new form…

But if the thought or if the knowledge or the new identity which enters in that conscious mind of that conscious being is a limited thought or is a limited knowledge or is a limited identity, thus that thought or knowledge or that identity, even though some type of energy or be it negative or be it positive, does not take that conscious being very far or into a greater state of identity and that thought or knowledge or that identity will disperse and the conscious mind becomes once again as if empty.

And if that conscious being nothing does with his conscious mind to have thought or knowledge or identity so that the thought or the knowledge or the identity takes him to a greater state or to a greater identity in where not only his conscious mind will be refreshed or becomes as if a new mind but also his physical form or body also will refresh or even could be cured from certain lacks or faults, such as of that of deafness in one ear or both and also some emotional lack or fault such as loneliness or shyness...

But if the conscious being in the course of his life does not enter into a greater state of thought or of knowledge or of identity, thus the conscious being keeps on dying until he completely dies and his body will discompose until it turns to dust and the conscious being has lost his the very grandiose opportunity of rebirth and of continuing with life as if new in complete or in perfect harmony and also in complete or in perfect abundance, perfect because it will be an abundance which will never ever end...

Now then, once the universe stops from discomposing or no longer the vacuum of space ever returns to nothing because of the conscious being coming to their maximum state or coming to their maximum identity and that way keeping the universe practically alive, thus there no longer will enter more knowledge in the form of matter or in the form of stars or of light because now that makes it possible the conscious beings because they will be the matter or the stars or the light or the illumination of the universe because of they being illuminated until the maximum or until perfection...

In other words, there will no longer be any more beginnings nor there will be anymore ends or anymore endings and the universe will become as if there were never beginning because of the universe becoming as if new for all of eternity...

And all the different parts of existence will act or will react as if one as the same the body and the conscious mind of the conscious being will act or will react as if one or as a single part and existence will be one with the conscious being because of the conscious being becoming or as being existence herself and reflecting through his body her glory...

Thus, when one as seed for more came out from the entrails of a man and one entered into the entrails of a woman, one had no memory of that exit or entrance even though one as a seed was in harmony and in abundance in those entrails and one also came out with all gladness and with all joy and also with all feeling of abundance and entered into the entrails of a woman and there also sought for knowledge of life to life receive and in her also enter and once in her also one forgot because once again one entered in harmony while one was transformed or one took the form of life which one did for or for the one which one received through the act of one or because of one's movement to find life...

And when one became complete in those new entrails, one humbled and one took the grandiose position of contender and one came out or one entered into the entrails of the world not only as much more but also one came out or one entered for much more.

But in the world one did not remember that one came out from harmony while one keeps completing the form of contender which could take one to not only come to be conscious but also which could take one to rebirth and continue with life as if with a new form or body.

But if there were no rebirth because of lack of knowledge or because of lack of identity, thus that form not reborn would take one to death and that would be the end or the ending of all of her, life, and also of all of one…

More about Existence

The more understanding of existence or of reality one has of reality or of existence, therefore, the greater existence or reality will be to one as one will be to reality or to existence.

Reality or existence or the universe has to do with knowledge and acknowledgement, even unseen reality has to do with knowledge and acknowledgement even if negative knowledge or acknowledgement.

Space is an unseen reality or unseen knowledge or acknowledgement but space can be acknowledged but only when there is knowledge to acknowledge space with.

In other words, space can really be acknowledged through or because of matter or light.

And matter as well as space can be divided into three major parts. Space is not only emptiness or is lack but space is also dark and cold.

These three parts which make space are lack or are negative (-) knowledge or negative (-) acknowledgement.

Positive knowledge or acknowledgement, therefore, is matter and its three major parts, such as weight or mass, light and heat.

And matter only interacts with space while in the vacuum or emptiness of space.

Now then, existence is about knowledge and acknowledgement or confirmation and reconfirmation or on and off or 01.

When matter which is knowledge enters the vacuum of space, space begins to react with matter as a form of acknowledgement and the vacuum of space begins to compress matter until matter explodes and it begins to expand in the vacuum of space.

But the friction which matters has from space makes matter to last less in the vacuum of space.

Now, before matter entered the vacuum of space, space was in a state of tranquility or space was neutral or 0.

But when matter entered the vacuum of space, space acted as if negative due to the interaction with matter such as the compression of matter but when matter exploded and began to expand through the vacuum of space, space began to act as if positive.

But when matter begins to turn off due to lack of energy, space also begins to act negative.

And when black holes begin to appear due to the collapse of large stars, the vacuum of space begins to act even more negative.

And when there no longer is matter in the vacuum of space because the black holes vacuumed all the matter up, the vacuum of space has become completely negative due to the black holes now expanding throughout the vacuum of space in search of matter.

But just as matter ran out of energy, so will the black holes run out of matter and collapse or disperse and the vacuum of space will once again become neutral or return to 0 and once again become ready to receive matter or new knowledge or 1.

But this new beginning is as if the very first beginning because there will not be any memory that there ever was any other beginning or beginnings.

Also, the vacuum of space is dimensional and will receive matter or knowledge or 1 up 118 times.

That is to say, there is 118 beginnings which really begun at the same time or spontaneously and enter into 118 dimensions or space time.

But not only that, matter or knowledge or 1 enters the first dimension or space time as 118 percent and that 118 percent equals or adds back to one.

In the same manner, 2 equals 236 percent and that 236 percent equals or adds back to one and so forth with the other numbers up to 118. That is, 3 equals 354 percent and that 354 percent equals or adds back to one.

And finally, 118 or the last number equals 13924 percent, which is 118 times 118, and that 13924 percent equals or adds back to one.

The dimensions of time also add up as if they were one. And they increase in size from one to 118 times.

That is, the first is one but the second is twice as the first and the third is three times as the first but they still all add up as if they were just one dimension.

And if the speed of light was ever measured correctly the speed of light would be equal to one. The speed of light is about 186, 282 miles per second which adds up to 9 or point nine or 09. This loss of 1 or of 01 is due to the compression of the vacuum of space. It is similar to water when it freezes. When water freezes water loses energy and weights from 1 to point 9.

Thus, existence or reality is one or existence or reality is knowledge or more like 01 which really means knowledge or acknowledgement.

For the conscious being, therefore, this information is very important because for this information the conscious being unknowingly struggles or contends for to be able to enter into higher or taller or greater conscious modes or be truly illuminated and through that illumination enter into a higher or taller or greater conscious existence and with this greater conscious existence or identity the conscious being can mode or transform existence or reality or his natural environment according to his will.

Of course, that forming or transformation really is done through his mouth and through the use of the spoken word or the zero plus one factor or language, more like binary because of the presentation or the knowledge and the acknowledgement.

Chapter One

The Universe

((((((((-) (0) (+))))))))

What is space?

What, then, is space?

Space is the absence of a matter, light or energy.

Space is the absence of a physical or of a solid body or the physical absence of a solid or of a physical identity.

Space also has no mass or space has no density or weight.

Space is, therefore, non-existence or nothing!

Space is also a vacuum or emptiness.

Space, is therefore, the absence of a physical reality.

That is to really say, space is nothing, space is empty and space is lacking.

Space is measured, however, not by the amount of space or the amount of emptiness but by the physical size of the actual or the real thing or the physical or the solid matter, light or energy or even heat that is going to actually fit or actually occupy that space or that emptiness.

So in reality, space in the physical sense simply does not exist or space is non-existence!

(((((((-) (0) (+)))))))

What is darkness?

Darkness is also the physical absence or the lack of light or Darkness is also the physical absence of energy.

Darkness also has no mass or Darkness has no density.

And what is cold or coldness?

Cold or coldness is also the physical absence or the lack of heat.

Cold or coldness is the absence or is the lack energy or heat.

Cold or coldness also has no mass or cold or coldness really has no density.

((((((((-) (1) (+))))))))

What is matter?

Matter is that which occupies space and matter is also that which has mass or volume.

Matter is that which is physical or is solid.

Matter is that that has mass and matter is also that which has density.

Matter is also physically affected by motion or by physical movement or by time.

Matter also physically exists or matter simply but physically is.

Matter, is therefore, physical reality.

Matter simply is physical knowledge of existence or actual reality or existence!

That is to really say, matter is also a product of physical reality or physical creation.

In fact, if no matter, then no physical reality, no universe or no physical creation!

$$((((((((-)\ (1)\ (+))))))))$$

What is light?

Light is that which occupies space. Light also has mass and light also has density. Light also physically exists or light simply but is.

Also, light is a physical product of physical reality or of physical creation.

$$(((((((-)\ (1)\ (+)))))))$$

What is heat?

Heat is that which occupies space. Heat also has mass and heat also has density. Heat also physically exists or heat simply but is.

Also, heat is a physical product of physical reality or of physical creation.

Thus, darkness is to space as light is to energy or to matter!

That is to say, darkness is to space or to cold or to coldness as matter is to energy or to light!

In short, space is to vacuum or to emptiness, to darkness and to coldness as matter is to motion or to expansion, to energy and to light.

Also, space is to nothing or space is to nothingness as matter is to something or is to physical reality.

In reality, therefore, space is to complete emptiness or space is to unseen reality as matter is to physical fullness or to physical reality.

Now let's simply say once again that space, darkness and coldness are simply but the very same thing.

And let's say also once again that matter, light and heat or energy are simply but the very same thing.

Now, there are two very simple opposites to work with or to work in: Nothing and something, or space and matter, or Non-existence and physical Existence, or Negative and Positive, or Black and White.

Or simply, that which is and that which is but not!

Space, coldness or darkness and matter, energy or light actually or really form or create the universe or physical creation, which is round or sphere-like, where space, coldness or darkness and matter, energy or light simply unite, physically join or simply intersect or physically meet or even divide.

Now, what is the solid or the physical thing with mass or with density that makes but a simple sphere or even a round shield?

What makes but a field or a sphere or even a simple shield?

What two figures or what two objects or what two things make a simple sphere or a simple circle where the two things physically unite, physically meet, physically interact or physically join or even divide?

What really makes but a sphere or shield?

What really makes a round field?

Or what is the very thing that makes a field?

What about two magnets?

What about two magnets in the form of cubes?

Yes! Where the two magnets or where the two magnetic cubes, this case matter and space, physically join, physically react or interact or physically unite or even divide, at the corners, the two magnetic cubes or the two magnets actually or really form a simple sphere or a field or a shield.

In fact, many or countless spheres, fields or shields or many or countless circles are physically formed by the two magnets or by the two infinite magnetic cubes!

But how are the magnetic spheres or the magnetic circles or the magnetic field or magnetic shield formed?

The magnetic spheres or the magnetic circles are formed due to the magnetic fields or magnetic shields of each of the magnetic cubes.

An excellent or a perfect and the most simple and the most unique example is the magnetic field or the shield of a bar magnet, **(-(-(-(-(-(-(0)+)+)+)+)+)+)**.

The bar magnet really creates or makes a special physical condition in the empty space or in the emptiness around the bar magnet.

That space or that special physical condition around the bar magnet is really noted as a magnetic field or as a magnetic sphere or as a magnetic shield.

The magnetic field or shield can also be simply called a field or shield of physical attraction.

The two cubes' physical or magnetic attraction for the other cube makes possible the magnetic spheres or makes possible the magnetic circles.

That is to simply say, instead of the magnetic field or shield or the magnetic sphere physically going around each of the cubes as the magnetic field or shield does with a bar magnet, the magnetic field or the magnetic sphere simply turns toward the opposite magnetic cube, thus making really possible the magnetic field or shield or the magnetic spheres or the magnetic circles or even the magnetic rings.

To simply illustrate or to simply show the magnetic field or shield, the magnetic spheres or the magnetic circles of the two infinite magnets physically attracting or physically interacting with one another, the Grid or the Rectangular Coordinate System is used, which is the Cartesian coordinate system, **(((((((((-) (0) (+)))))))))**.

Only two sections or only two quadrants, sections or quadrants number two and sections or quadrants number four of the Grid or the Cartesian coordinate system are used.

Quadrant number one and quadrant number three can also be used to show or to illustrate the two infinite magnets or the two simple magnetic cubes.

The simple reason for this selection of the two sections or the two quadrants is that the selected two sections or the two quadrants physically meet or physically interact only at one physical point.

And that physical point really is physical point zero, physical point of impact, physical point of intersection or physical point of interaction or physical point of origin or even physical point of view, **(((((((((-) (0) (+)))))))))**.

It can also be simply said that the two selected sections or quadrants are physically opposite to one another or facing one another.

As it can be simply seen from the diagram or the simple drawing, **(((((((((-) (0) (+)))))))))**, where the two magnetic cubes or where the two magnets physically meet or even divide, many magnetic circles or many magnetic spheres are physically formed!

But the magnetic circles or the magnetic rings or the magnetic fields or magnetic shields must be sphere-like, for they are really the product of two magnetic cubes.

In reality, countless numbers of magnetic circles or countless numbers of magnetic spheres are formed for the Cartesian coordinate system goes on forever and ever.

Also, the magnetic circles or the magnetic spheres or the magnetic rings increase in size as real or as whole numbers increase in size by one.

That is to simply say, the size of the first magnetic sphere or magnetic ring is one or is number one.

The second magnetic sphere or magnetic ring is two or twice the size as the first magnetic sphere or magnetic ring.

And the third magnetic sphere or magnetic ring is three times the size of the first magnetic sphere or the first magnetic ring, and so on and so on.

Now, inside one of those magnetic spheres or inside one of those magnetic circles really is the universe!

When the two magnets, in this case the space cube and the matter cube or the light cube, physically attract one another through their magnetic fields, the two magnetic cubes create many magnetic spheres, $((((((((-) (0) (+))))))))$.

Inside one of those magnetic spheres is the universe, $((((((((-) (U) (+))))))))$.

But what, therefore, makes possible the magnetic cube?

What makes possible the cube?

That is to say, how does the cube gets its unique shape?

Since the infinite magnetic cube really is outside of space or is outside any vacuum of space or is outside any gravity or is outside any compression, there really is no compression or is no vacuum to physically act or to physically interact upon the infinite magnetic cube, thus really allowing the infinite magnetic cube to expand throughout and, therefore, making or taking its unique cube shape.

Now there are two infinite magnetic cubes and, where they physically meet, physically interact or physically intersect, countless magnetic spheres or countless magnetic circles are formed; thus, the physical universe.

One of those magnetic spheres or one of those magnetic circles is the physical universe.

Actually, every magnetic sphere or every magnetic circle is really a separate universe or a separate dimension of time!

Both infinite magnetic cubes physically attract one another or one cube physically attracts the other cube with the very same magnetic force or magnetic attraction.

That is to say, one magnetic cube attracts or physically pulls the other magnetic cube with equal magnetic force or equal magnetic attraction.

The infinite matter, the infinite light or the infinite energy cube physically attracts the infinite space, the infinite coldness or the infinite darkness cube, but the infinite space cube releases nothing for infinite space, being empty, has nothing to release.

The infinite space, the infinite coldness or the infinite darkness cube physically attracts the infinite matter, the infinite energy or the infinite light cube, which has density or is solid or is physical, and a physical part of the infinite matter, the infinite energy or the infinite light cube is broken off or is released, and takes up space in the magnetic sphere or takes up space in the magnetic circle formed in the universe or the vacuum of space by the two infinite magnetic cubes, $(((((((-) (1) (+)))))))$.

Once the solid or the solid part or the solid piece of matter which is in the physical form of a smaller cube of matter, or simply call it the universe's solid egg, from the infinite matter, the infinite energy or the infinite light cube, physically enters the vacuum or the compression of space, there is the actual or the real and the physical beginning or the physical laying of the universe's egg.

Since empty space is super cold and also a super or a giant vacuum, the physical properties of matter, energy or light, or the smaller matter, energy or light cube as an egg in the vacuum of space, physically changes.

Matter, energy or light now in the super vacuum or in the super compression of space shrinks or matter, energy or light is super compressed in the super vacuum of space.

Once super compressed, matter, energy or light as a smaller cube is now super compressed into a sphere of matter, energy or light.

The new sphere of matter, energy or light in the super vacuum of space super heats and super explodes and thus simply expands throughout the universe or simply expands throughout the simple magnetic sphere or simply expands throughout the magnetic circle or magnetic field or magnetic shield.

Thus, the Big Bang!

Actually, the Big Bangs, all 118 of them!

$$(((((((-) \ (1) \ (+)))))))$$

As it can be really be seen

As it can be really be seen, the rest of the universe or the rest of actual or real existence or even creation is time, physical evolution, history and simple ideas, true or not.

That is to simply say, the rest of the real or the actual universe or the rest of the real or actual reality or physical existence or physical creation is simply the Big Bangs of human ideas!

The reality of the matter is that there are really two creations.

There is physical creation or physical reality.

And there is mental creation, but mental creation or human thought does not let physical creation or actual reality thus be seen, and much less enjoyed.

Just because it is believed, for example, it does not make it so.

The truth or reality is not something one just believes in, but the truth or reality is something that can really be seen.

When mankind does not understand something or even see it, mankind humanizes it or makes it unnatural.

The universe or existence or physical creation is as the conscious living being simply is.

The more the conscious living being simply is or the more the conscious living being simply becomes or really knows or understands, the more the universe or the more physical or the more actual existence or the more physical reality or the more physical creation simply will be!

In other words, the universe or existence herself or even physical creation is only limited to the conscious thought or only limited to the conscious belief of the conscious living being.

Limit the conscious knowledge of the universe or limit the knowledge of actual existence or of actual physical reality or of physical creation, thus limit the conscious living being.

Existence, the universe or physical creation is composed or really made up of three main parts.

The three main parts are physical reality, neutral reality or point of view, and the reality not seen but can be negative or irrational.

Or simply put, all of Existence is "(+) (0) (-)."

In other words, $E = +0-$. That is, plus zero minus.

The same formula can also be used for the universe.

That is to really say, "$U = +0-$."

Thus, the universe is really equal to the actual reality recognized plus personal point of view or personal understanding or the reality actually seen minus the reality not yet recognized or the reality believed.

The greater the plus is the greater the point of view, and the lesser the negative or the lesser the irrational.

The greater the negative or the more the irrational, the less the point of view or the less of actual physical creation is really seen and, therefore, the less understood and the less enjoyed or the less acknowledged.

But if the understanding or if the enjoyment or if the acknowledgement is greater or taller, thus greater or taller is the universe or thus greater or taller is existence.

(((((((((((((U)))))))))))))

Dimensions of time

In reality, there is a finite or there are only 118 separate universes.

In reality, there is a finite or there are only 118 dimensions of time between the two infinite magnetic cubes: the space, the cold or the darkness cube, and the matter, the energy or the light cube.

Even though the two infinite cubes are encircled in magnetic spheres, the first 118 magnetic spheres are actual universes or are actual dimension of times.

The first 118 magnetic spheres are actually the middle, the center or the belly bottom of existence or of reality or physical creation.

Every magnetic sphere, for example, illustrates a separate universe or a separate dimension of time, (((((((0))))))).

The number of universes or the number of dimensions of time, once again, is in reality finite or the number is limit even though the magnetic spheres are countless as the numbers are infinite or countless.

What this really means, is a magnetic sphere inside a magnetic sphere inside a magnetic sphere until infinity, (((((((0)))))))...

However, only the first 118 magnetic spheres are actual universes or are actual dimensions of time.

Each magnetic sphere or every universe is separated by a dimension of time or is separated by a dimension of space.

The magnetic circles or the magnetic spheres increase in size as the two infinite magnetic cubes extend forever or extend until infinity.

The motion, the movement or the expansion in Time or in the universe or the expansion of the universes is created or is caused by the magnetic field of the two infinite magnetic cubes.

By the way, not every magnetic sphere is a dimension of time and not every magnetic sphere is a separate universe.

The dimensions of time are finite as the universes are finite in number. There are only 118 dimensions of time as there are also only 118 separate universes.

When empty space or the vacuum of space attracts the infinite matter, the infinite energy or the infinite light cube, not only does the infinite matter, the infinite energy or the infinite light cube releases a smaller matter, energy or light cube with 118 smaller matter, energy or light cubes, but also the infinite matter, the infinite energy or the infinite light cube releases multiple matter, energy or light cubes at the very same time.

The infinite matter, the infinite energy or the infinite light cube releases smaller matter, energy or light cubes with 118 smaller cubes each according to the size of the magnetic spheres or the size of the universes.

That is to say, the infinite matter, the infinite energy or the infinite light cube releases at the very same time one matter, one energy or one light cube with 118 pieces of matter, energy or light for the number one sphere or the number one or the very first universe; the infinite matter; the infinite energy or the infinite light cube releases two matter, energy or light cubes with 236 pieces of matter, energy or light for the number two or the second sphere or the number two or the second universe; the infinite matter, the infinite energy or the infinite light cube releases three matter, energy or light cubes with 354 pieces of matter, energy of light for the number three or the third sphere or releases three matter, energy or light cubes for the number three or the third universe, and so on until the very last universe or on until the very last dimension of time, which is dimension number 118.

Dimension number 118 or universe number 118 is 118 times the size of dimension number one or 118 times the size of universe number one.

In other words, magnetic sphere number118 is 118 times the size of magnetic sphere number one. Universe number 118 is really 118 times the size of universe number one.

For every one cube of pure matter, energy or light, there are 118 smaller matter, energy or light cubes or 118 pieces of matter making that one matter, energy or light cube.

The very same number of matter, energy or light that enters the vacuum of space is the very same number of space taken or occupied by the matter, the energy or the light.

As soon as matter, energy or light from the infinite matter, the infinite energy or the infinite light cube enters empty space, cold or darkness or matter enters the vacuum of space or enters the magnetic spheres created by the magnetic field of both the infinite matter cube and the infinite space cube, matter, energy or light in the vacuum of space is super compressed due to the enormous pressure or due to the enormous vacuum of space, cold or darkness.

The enormous pressure or the enormous vacuum in space causes matter, energy or light in the vacuum of space to super compress into a giant or into an enormous magnetic sphere of matter, energy or light in the vacuum of space and to superheat and to super explode, thus making possible the Big Bang!

Actually, making possible the Big Bangs! Or the hatching or the hatchings!

Enormous energy is loss from the super Big Bangs! Or from the super hatchings!

But a lot of matter, energy or light is still left in the vacuum of space to spread or to expand throughout the vacuum of space and spread or expand throughout the universes or throughout the magnetic spheres.

Some matter will lose energy, therefore, becoming stars, planets and moons and some matter actually becoming even smaller pieces or bits of matter, such as space rocks or space dust.

The heaviest or the denser matter or the heaviest or the denser stars at one point in time or at one point in the vacuum of space will turn into supernovas and then implode into super black holes or turn into the cancer of the universe and universes.

Actually, black holes or collapsed stars are like giant vacuum cleaners or giant space sweepers, really giant magnetic graves which will destroy all matter, energy or light!

There will be another point in time or another point in the vacuum of space when most of the universe and universes will be occupied by super or by giant black holes or by giant cancerous stars or dark stars or vast empty magnetic graves.

Once again, the universe or universes will be occupied by giant vacuum cleaners or giant sweepers or giant graves.

The universe and universes will look like a giant Swiss cheese, but without the appetizing smell.

At that very stage, when the entire universe and all universes are occupied by enormous black holes or enormous dark stars or enormous vacuum clears, the enormous black holes or the enormous dark stars will disintegrate or will drain the entire light or the entire energy or the entire matter of the universe and universes spontaneously or at once!

That is to say, the giant black holes or the giant space vacuum clears will begin contracting or moving or going to the center of the vacuum of space or to the actual physical point of impact or to point zero or to the point of origin until the entire physical universe and the other universes collapse—the exact opposite of the Big Bang! Or Big Bangs!

The black holes or the collapsed stars will begin to move toward the center or begin moving toward the actual physical beginning or the physical origin of the universe and universes, removing all the matter, all the energy or all the light from the entire universe and universes and thus making one giant or making one enormous black hole or one giant or one enormous vacuum per universe or one enormous vacuum per magnetic sphere.

It is this new giant or this enormous black hole vacuum in the vacuum of space or inside the magnetic sphere that actually attracts new matter, new energy or new light from the infinite matter, the infinite energy or the infinite light cube.

Once new matter, new energy or new light from the infinite matter, the infinite energy or the infinite light cube enters or re-enters the giant black hole vacuum in the vacuum of space or in the magnetic sphere, the new matter, the new energy or the new light is super compressed recreating the Big Bang! Or really creating the Big Bangs! And destroying the giant or the enormous black hole vacuum!

Actually, every giant or every enormous black hole in the vacuum of space or in every occupied magnetic sphere or in every occupied dimension of time or in every occupied dimension of space will be destroyed at once or at the very same time by new matter from the infinite matter, the infinite energy or the infinite light cube!

In other words, every giant black hole vacuum occupying the 118 universes or occupying the 118 dimensions of time will be destroyed at the very same time by new matter, new energy or new light from the infinite matter, the infinite energy or the infinite light cube.

However, there also exists the possibility that the giant black hole or the black holes will disperse as if a storm and space will become as if a sea of tranquility.

Chapter Two

The Stars and the Black Holes

The universe is a neutral point or is a point zero which is converted into a positive point or into a one or more when light or matter enters into the universe.

The universe also is converted into a negative point when the grand majority of light turns off or the majority of matter no longer has energy…

Before light or matter enters into the universe the weight of the universe is cero and when light or matter enters into the universe thus the universe takes on weight.

Light or matter is composed of 118 Elements and the weight of each element is two times its atomic number, more like its positive number.

In the case of element number one, for example, its weight is of two and in the case of element number two its weight is of four and in the case of element number 118 its weight is of 236…

Curiously, that the totality of the 118 elements adds to one and that the totality of their weight adds to 2.

That is to really say, that if we added from one to 118 thus the sum would be 7,021.

And if we added that sum thus it would be 10 and if we add that last sum thus it would be one. That is to really say, 118 is equal to one…

And if we added from two to 236 the sum would be 14, 042.

And if we added that sum thus it would be 11 and if we added that last sum thus it would be 2. That is to really say, 236 are equal to two…

Thus, light or matter enters into the vacuum of space or into the universe as one or as a unit which is composed of 118 pieces or elements and the weight of the element is two times the atomic number of the element.

Thus in truth, number one itself is composed of 100 percent plus 18!

That is to say, that the number one or even oneself is equal to 118 percent!

Also light or matter could enter into the vacuum of space or into the universe with only or as one element with its weight of two but that element would be able to convert into the other elements, even to the element 118 and its double weight of 236…

Element number one, for example, enters into the vacuum of space with its weight of two and it has 117 other possibilities of converting into the other 117 elements.

That is to say, element one is composed of 118 parts or pieces or the 118 percent and element number has 117 possibilities of converting into the other 117 elements according to the weight which element number one maintains.

In the same manner, element number two with its weight of four has 116 possibilities of converting into the other 116 elements or until the element number 118 with its weight of 236...

When light or an element enters into the vacuum of space, light or the element enters as if it were a piece of magnet or as if it were a bar magnet.

In the vacuum of space the magnet or light or the element is super compressed not only until it takes the form of a sphere or round but also light or the element or the magnet is super compressed until it gets to a very high level of temperature.

And when the temperature gets to its highest level thus light or the magnet or the element super explodes causing the light or the magnet or the element to divide into two parts and the part with less weight, such as the negative part, takes position in the magnetic field and that magnetic field now is a negative field...

The superior or the positive or the heaviest part of light or of the magnet or of the element takes position in the center or in the nucleus.

Thus, now we have the negative part of the magnet going around the positive part when before they were united and the center or the nucleus was neutral...

And even though the element lost weight because of the explosion or because of bursting, the element continues the one for two.

That is, its weight continues of two although the element one now is 0.999 and its weight is double, 1.998...

Thus, now element number one was reduced to about 0.999 with its new weight of 1.998 but to element number one also remains a neutron or even more than one or a neutral part which can be converted or can be transformed into a positive part and that way not only adding the number of the element but also adding its weight even though it will have only one negative particle going around the center or the nucleus or the positive side...

But if one or light or the magnet or the element does not convert into the next number or into the next element thus it loses its energy and will only be a piece of dead matter in the vacuum of space and it will be removed one day by the black holes...

Thus, light or the element is the same as a negative particle, is the same as a neutral particle and is the same as a positive particle.

In a way, one is equal to a negative portion plus a neutral portion plus a positive portion which totality is of 0.999 after entering the vacuum of space.

But in the vacuum of space light or the element is positive even though the vacuum of space is neutral but reacts as if negative because of the vacuum.

And when the element increases its positive part by converting the neutral part into a positive part, the element cannot attract the negative part because of the vacuum of space because now the negative part becomes as if more or its weight increases because of the weight that it receives indirectly from the vacuum of space.

And if there were not positive attracting the negative thus the negative would expand through the vacuum of space and it would stop being negative and it would be dead matter...

Now, an element, in this case a star, which number is high as the same as its weight thus lasts or remains longer in the vacuum of space or in the universe.

But the element or the star becomes heavier while the energy or matter to continue on lasts or it begins to transform from positive to neutral and once neutral, the element or the star practically becomes negative when its excess or super weight attracts the electrons toward the center and thus causing an implosion in where the element or the star becomes a nova or a new star but without light or without energy and that way causing an enormous hole in the vacuum of space when before the element or the star occupied the vacuum of space as an element or as a star...

In other words, matter or the star in the vacuum of space changes from positive to neutral and then from neutral to negative.

Thus, + 0 -, in where the negative is a black hole or a super vacuum cleaner in the vacuum of space.

This black hole practically eats or sucks all the matter around it to take all matter out from the vacuum of space or from the universe to make new space for new matter or for another beginning...

But as long as there are black holes in the vacuum of space or in the universe, thus the vacuum of space or the universe is negative and as long as the vacuum of space or the universe continues as or is negative thus it keeps being for something and not for neutral and from neutral or from zero to positive...

Thus, so that the universe becomes neutral or to zero and from neutral or from zero to positive thus all the black holes or super space vacuums must stop from functioning and once the black holes or the super space vacuums stop from functioning for lack of matter or for lack of energy thus the vacuum of space or the universe will return to neutral or to zero…

Now then, this new vacuum in space or in the universe is neutral and has zero energy but even so attracts new matter or new stars or attracts the new or the next beginning which is outside of the universe or outside the vacuum of space…

And once new matter or new stars enter into the vacuum of space or into the universe thus the vacuum or the universe will be positive or one or more…

Thus, the cycle or the model or the rhythm of (+ 0 -) continues until the end of all the times…

The Universe and the Number

The Universe as is Creation truly is about giving name or naming to start or to begin and giving rename or renaming to continue on without beginning again even though to give rename or renaming is to become as if forever new and as if never there were a start or a beginning and neither an end...

The Universe as is all of existence truly is composed of numbers or be it positive numbers or be it neutral numbers or be it negative numbers or be it a combination of all the numbers at the very same time.

But there will always be a number that is greater than the other negative numbers until that positive number is converted into a neutral number and after that positive number is converted into a neutral number thus also it will be converted with time into a negative number or a void.

In other words, the Universes as is existence really is equal to $(+ 0 -)$, but existence does not return to negative but rather a part of her and that part of her is the universe or matter or light, which returns to or truly is converted from positive to neutral or to zero and much later it is converted to negative or into nothing.

The number zero is not negative or nothing or void. The number zero only is a neutral number which could be converted into a positive or negative number.

Existence as matter or as light or energy is truly composed of three main parts which add to 118 percent, 118 times 3 because they truly are 118 positive parts, 118 neutral parts and 118 negative parts or (+ 0 -).

Now then, the neutral part or part zero is the universe and it is where it is added, it is subtracted, it is multiplied and it is divided at the very same time.

The positive part is the physical part of existence or the part from where comes out matter or light or the elements which truly are pure matter, but this part of existence appear to be minor than the other parts of existence.

The negative part of existence is the lack or is the vacuum of space or darkness which is composed of three parts, one is space or emptiness and the other two are space-time, which truly are created by the interaction of the magnetic field between the positive and the negative.

When matter from the physical side of existence enters into the vacuum of space, matter even though a single piece, matter enters with its 118 parts assimilating the 118 elements.

That is to say, if only one element enters into universe or the vacuum of space that element could assimilate or even can come to be converted into the other 117 elements...

But in reality matter enters from the physical side of existence into the universe or the vacuum of space in 118 pieces into 118 parts of the universe or the vacuum of space or into 118 dimensions at the very same time.

That is to say, into the universe or into the vacuum of space there enters 118 elements and every element has 118 parts and at the very same time there enters 118 elements into 118 dimensions.

But that, however, does not remain like that, because with every element the 118 parts are also multiplied.

In other words, element number one has 118 parts but element number two has 236 parts and element number three has 354 parts and this pattern continues on until the last element, which is element number118.

Also, the weight of the element is twice its atomic number. That is to say, element number one has a weight of two and element number two has a weight of four and element number three has a weight of six. This pattern also continues on until element number 118.

Interestingly, however, that all the parts of an element add to the atomic number of that element!

Thus, element number one has 118 parts and if we added those 118 parts thus we would get one. That is, if we added 1+1+8 thus they would give us10 and if we added 1+0 thus we would get 1.

In the case of element number two, if we added 2+3+6 thus we would get 11 and if we add 1+1 we would get 2. This pattern continues on until element number 118. Element number 118 has 118 parts and every part also has 118 parts thus giving a sum of 13,924, also adding to one.

Now then, when element number one enters into the vacuum of space element number one enters with one proton, with one neutron and with one electron, (1+0-1). And also element number one enters with a weight of two.

But once in the vacuum of space, space comprises the element and the element is comprised causing the element to super heat and thus also causing the element to burst or explode.

In that burst or explosion the element loses a neutron or a third part and no longer the weight is of two but is less, such as 1.67.

Also the electron was separated and now is spinning around the proton or the nucleus of the element, (+) -.

In the case in where an element has a high number, due to its prontons, thus that element will have neutrons in its nucleus, such as (+0+) -, -. Or (+0+0+) -, -, -.

This new transformation of the element, in where the electron is separated from the nucleus makes it possible for the element to be united to other elements and that way converting into a mixture of elements or isotopes…

Furthermore, before an element enters into the vacuum of space its three main parts actually have the same size or weight, ((+) (0) (-)).

But once that element enters into the vacuum of space thus that elements is divided into two parts, the nucleus in where now is the proton (+) and the neutron (0), and the outside part in where now is found the electron (-) going around the nucleus, ((+) (0)) (-).

And while the nucleus keeps its size or weight, the electron loses its size or weight because of the interaction or friction which it has with the nucleus or with the center of the element.

The interaction or friction which the electron has with the nucleus or with the center of the element also causes the electron to last less or lasts less time in the vacuum of space.

Once the electron is fused or is exhausted, the element or the nucleus is converted into a neutral element or without energy even though the nucleus is still positive or with protons and neutrons.

But the outside part or the electric field of the element is now a neutral field or the electrons have been converted into neutrons because of lack energy.

In other words, the seven electron rings of the element or the atom now are neutral when before they were negative because of the electron.

And just as the element functions thus that way also functions the number and existence herself, but the number or the symbol of the number is only an illustration of the numbers but truly does not show how is the number in existence or outside the universe or the vacuum of space in where there is no friction or movement even though there is a magnetic field.

Thus in truth, the number one or 1 outside the vacuum of space is represented by a cube.

Now, the cube or the number or the element one is composed of three main parts and they are the positive part, the neutral part and the negative part, (+) (0) (-).

But those parts also are composed of 118 other parts. That is to say, that the positive part also is composed of 118 parts and the neutral part is also composed of 118 parts as the same as the negative parts which also are composed of 118 parts.

And when the cube or the number or the element enters into the vacuum of space the cube or the number or the element is compressed into a sphere or into a globe but still with its 118 parts.

Thus, the number one is composed of not only 100 percent but also of another 18 parts or of another 18 percent.

And if we added the parts thus we would have 1.

Now then, the number one or the symbol 1 as also is all of creation is a continuation because the start or the beginning is zero or a point or a neutral or an empty space.

But the number one or the symbol 1 also represents all of existence, the physical part, the neutral part and also the negative part.

Also the number one has the ability of converting itself into its 117 other parts also with their other 118 parts.

In other words, the number or element one also has the ability of being infinite because also it could renew into a greater number such as the number two.

And that makes it possible the other 117 parts which add to 9 and the number 9 is a symbol of renovation.

That first renovation extends the time of the number or of element one.

Thus, if we added the 117 parts which remain to one plus its other two parts, the neutral with its 118 parts, and the negative with its 118 parts, the sum would be of 353.

And if we added 353 thus it would give us 11 or eleven and if we added 11 we would get two, the possibility or the ability of the number or element one if it is renewed.

And once that the number or that element number one has renewed as two, thus the number or element one has become or will continue as double or for much more as two and as double the abundance.

And the very same step or process is with the number or with element number two. If we added all the parts that the number or the element two has, which are the double of one, thus it would give us four or 4.

That is to say, if we added all the parts of the number or element two thus we would get the double.

The sum up the universe, the universe is finite but existence is infinite or without size.

The universe ends or stops running but existence runs or she expands toward all sides at the very same time and at the very same time existence also compresses toward all sides, that way also adding to her very self an infinite weight and an infinite size.

That is to say, existence adds to herself, she multiplies herself, she divides herself and existence also subtracts her very self or from her very self to be able to become forever for more and even for new.

The universe begins but existence forever existed and existence forever will exist. Existence has no time even though time existence is.

Existence is also composed of three principal parts which all add to one and that one is the grandiose part which makes the difference in all of existence.

Existence consists of matter, that which is physical. Existence also consists of space or of vacuum, that which is lack of something or lack of matter, energy or light.

And existence also consists of movement or of that which many call space-time, which is a movement as if in waves in space.

Thus in truth, existence is a simple magnet!

In other words, existence really is composed of +0-; in where (+) is equal to positive; in where (0) is equal to neutral; and in where (-) is equal to negative or the lack of.

In the scale of colors would be white, grey and black; in where white is light or matter; grey is neutral or space-time; and black is darkness or lack.

Now then, just as existence truly is composed of three parts and practically of parts opposite to the other parts, +0-, thus one also truly is composed of those very same opposite parts but in knowledge because one is knowledge.

And one can truly be positive knowledge or be neutral knowledge or even be negative knowledge.

But the greater the positive knowledge one has, less the neutral knowledge and less the negative knowledge or less the lack of negative knowledge.

In other words, the greater the knowledge of reality that one has, the greater the reality of one and less is the reality not known or less is emptiness or darkness.

Thus in truth, the more the knowledge of one, the more the abundance or the more is the life of one or more life makes sense to one.

Interestingly, that existence functions the same as one but existence knows it not but existence will never stop from being or will never stop from existing for all the eternity of eternity because she truly makes herself for much more!

That is in truth, even though all of existence is one, because of one existence truly is and for all eternity infinite!

And every time that existence is increased by one or is increased by more than one, the point of neutrality or the neutral is less as the same as the negative or the unknown point is less.

But making the things less does not make existence greater, but making the things or existence greater makes existence greater!

And just as existence makes the things greater for existence to be greater herself, oneself can be greater by recognizing that life is more or that in life herself there is more and that there is with all peace, the neutral, and that here is with all knowledge, one or more than one; and also that there is with all gladness and with all joy and that also there is with all abundance of life, life renewed...

Thus, through knowledge one renews for much more and because of one renewing for much more, one truly can continue for all of eternity as the very eternity...

Chapter Three

The Black Hole

Introduction

The Black Hole is a giant hole or a round rupture or tear or breakage in the fabric or the foreskin skin which really is the vacuum of space.

But the hole or the tear or the breakage in the vacuum of space moves along the vacuum of space sucking up or removing or destroying all matter, including all the stars and all planets and even all forms of life, to clear the universe or the vacuum of space for new matter or for another beginning which will appear as if the very first beginning because there will not be any trace that there was ever a first beginning...

But no need to worry because the above matter will happen billions of years in the future. By the time that our galaxy is intersected by a black hole for destruction, our sun has extinguished for lack of fuel or energy and life on our planet has also stopped from existing because of lack of illumination…

Thus, the true purpose of the black hole, however, is to turn the vacuum of space or the universe into a neutral point (0) to attract new matter or new energy from outside of the vacuum of space or from outside the universe which will become the new stars and the new planets of the new universe after the new big bang!

But as long as there is a single black hole in the vacuum of space or in the universe then the vacuum of space or the universe will be negative or negatively charged (-) and nothing will enter the vacuum of space or the universe as long as the vacuum of space or the universe remains negative or negatively charged…

The Super Black Hole

The heaviest or the denser matter or the heaviest or the denser stars at one point in time or at one point in the vacuum of space will turn into supernovas and then into super black holes or turn into the cancer cells of the universe and universes, yes there are other universes or other dimensions of times!

Actually, black holes or collapsed stars are like giant vacuum cleaners or giant space sweepers, really giant magnetic graves!

There will be another point in time or another point in the vacuum of space when most of the universe and universes will be occupied by super or by giant black holes or by giant cancerous stars or dark stars or vast empty magnetic graves.

Once again, the universe or universes will be occupied by giant vacuum cleaners or giant sweepers or giant graves.

The universe and universes will look like a giant Swiss cheese, but without the appetizing smell.

At that very stage, when the entire universe and all universes are occupied by enormous black holes or enormous dark stars or enormous vacuum clears, the enormous black holes or the enormous dark stars will disintegrate or will drain the entire light or the entire energy or the entire matter of the universe and universes spontaneously or at once!

That is to say, the giant black holes or the giant space vacuum clears will begin contracting or moving or going to the center of the vacuum of space or to the actual physical point of impact or to point zero or to the point of origin until the entire physical universe and the other universes collapse—the exact opposite of the Big Bang! Or the Big Bangs, all 118 of them!

The black holes or the collapsed stars will begin to move toward the center or begin moving toward the actual physical beginning or the physical origin of the universe and universes, removing all the matter, all the energy or all the light from the entire universe and universes and thus making one giant or making one enormous black hole or one giant or one enormous vacuum per universe.

It is this new giant or this enormous black hole vacuum in the vacuum of space or inside the magnetic sphere which is the universe that actually attracts new matter, new energy or new light from outside the universe or the vacuum of space.

Once new matter, new energy or new light from the outside of the universe or outside the vacuum of space enters or re-enters the giant black hole vacuum in the vacuum of space or in the in the universe, the new matter, the new energy or the new light is super compressed recreating the Big Bang! Or really creating the Big Bangs! And thus destroying the giant or the enormous black hole vacuum!

Actually, every giant or every enormous black hole in the vacuum of space or in every occupied universe or in every occupied dimension of time or in every occupied dimension of space will be destroyed at once or at the very same time by new matter from outside of the universe or from outside the vacuum of space!

In other words, every giant black hole vacuum occupying the 118 universes or occupying the 118 dimensions of time will be destroyed at the very same time by new matter, new energy or new light coming from outside the universe or outside the vacuum of space.

However, there also exists the possibility that the giant black hole or the black holes will disperse as if a storm and space will become as if a sea of tranquility or neutral (0).

Chapter Four

The Atom

[(+) (0) (-)]

The Atom is a microscopic magnet of matter or the element or even of existence, [(+) (0) (-)].

When the Atom enters the vacuum of space from outside of space in the form of matter or as an element, the Atom is super-compressed by the vacuum of space and thus creating a microscopic big bang, all 118 of them!

After the microscopic big bang the magnetic field of the atom becomes a negative field due to the electron which now has separated from the atom's center or nucleus and takes position on what used to be the magnetic field of the atom, [((-)) [(+) (0)] ((-))]...

Interestingly, even though the electron is speeding around the now electron field or ring, the electron also is composed of three main parts, but just as the atom is surrounded by a magnetic field or shield, the electron is also surrounded by an electron shield or shield or ring.

And so what we have here is negative going round negative going around negative, similar to our solar system in where the sun goes around the galaxy and the earth goes around the sun and the moon goes around the earth...

In other words, the electron also has its three main parts of [((-) (0) (+))].

When all of the electrons wear out because of the friction between the electron and the nucleus, the protons begin to react causing an explosion and thus draining matter or the element from all energy.

In other words, when there is no longer any negative part or electrons, the positive part or the protons alone with the neutrons collapse on themselves causing an implosion and thus the end of matter or the element because of lack of energy.

Now, matter or the element is dust or waste in the vacuum of space or in the universe...

And the end of matter or the element is the end of light or illumination and thus the end of the universe...

To sum up, the Atom really is a microscopic magnet or diagram of matter or the element or even of existence.

The Atom is also the identifying number or unique tag of matter or the element, which is the purest form of matter.

There are only 118 elements in the universe or existence and every element has its unique atomic number, more like its proton number. No two elements have the same atomic or proton number.

In fact, no two elements have the same atomic weight, which is the weight of the protons, the neutrons and the electrons.

Also, the weight of the atom or the element is twice is atomic or proton number.

Interestingly, that as the electron becomes smaller because of the friction with the nucleus the nucleus becomes heavier.

The above phenomena can be compared with the earth and the moon. The earth gets heavier and the moon gets smaller.

At one point in time the earth will pull the moon to the earth provide the earth gravity or the earth's magnetic field stays strong enough to pull the moon to the earth, creating vast destruction to both the earth and moon...

[(+) (0) (-)]

What is, therefore, the atom?

The atom is a microscopic or is a physical miniature of matter, light or energy.

The atom is a microscopic or is a physical miniature of existence or of the universe or of physical creation.

The atom is a microscopic magnet inside a magnetic sphere or in a magnetic field composed of seven electron or seven magnetic rings.

The atom as microscopic matter is also composed or made up of three main physical parts. The three main physical parts are protons, neutrons and electrons or simply put, "+0-."

That is to say, A=+0-. In other words, the atom is plus zero minus, [(+) (0) (-)].

Atoms are also transitional or turn from positive to neutral to negative, [(+) (0) (-)].

Contrary to its name, the atom is physically divisible or is physically broken into three main physical parts.

And those three main physical parts can also be divided or can also be broken into many smaller physical and identical parts as the first three main physical parts.

The atom is also made of protons (+), neutrons (0) and electrons (-). The seven electron or the seven magnetic rings of the atom are created by the atom's magnetic field or shield or the atom's magnetic sphere.

Actually, it is an atomic magnetic shield or an invisible magnetic or electron shell which are also called electron rings/

When the atom or when pure matter or when the element physically enters the vacuum of empty space, the atom is physically divided or physically broken or is physically separated from the atom's center or from the atom's neutrally charged nuclei or neutrally charged center, and the electrons occupy a specific physical part or a specific physical location or position of the magnetic field or the now electron ring.

What this actually means is that the atom went through a microscopic or a miniature physical expansion or the atom went through a microscopic or a miniature big bang!

Now, however, the first atomic number or the first element or element number one is the element Hydrogen. H is the symbol for the element Hydrogen.

Hydrogen has one proton or has one positively charged atomic or has one microscopic particle or has one atomic matter in the center or in the nucleus or nuclei of the atom.

The element Hydrogen also has one neutron or has one electrically neutral atomic matter particle in the nucleus.

The element Hydrogen also has one electron or has one negatively charged atomic or has one microscopic particle or has one atomic matter on the first electron or the first magnetic ring.

However, before Hydrogen enters the vacuum of space, Hydrogen really has two neutrons or two electrically neutral particles of pure matter.

When the element Hydrogen enters the vacuum of space, the element Hydrogen really losses one neutron as pure energy!

In fact, all matter or all elements or all atoms lost neutrons as pure energy in the vacuum of space.

Pure matter, the elements or the atoms have twice the number of neutrons before pure matter or the atom enters the vacuum of space.

For every one proton or for every one electron that enters the vacuum of space, one neutron is lost as pure energy in the form of heat. One neutron for each atom was lost. Pure matter, the elements or the atoms lost neutrons according to the atomic number.

For every proton or for every electron, one neutron is lost as pure energy in the vacuum of space. The loss of neutrons as pure energy in the vacuum of space is 1, 2, 3, and 4, and so on until the very last atom or the very last element in the Periodic Table of the Elements.

The very last atom or the very last element is atom or is element number 118. So, element or pure matter number 118 in the vacuum of space lost 118 neutrons as pure energy.

Helium, the second element on the Periodic Table of the Elements has two physical parts of atomic matter. The symbol for Helium is He.

What this really means is that Helium has two atoms or has two atomic particles of pure matter. That is to say, the two parts or particles in Helium means that Helium has two protons or has two positively charged pieces of pure matter.

Helium has also two neutrons or has two electrically neutral particles of pure matter in the center or in the nuclei.

For every one proton there is a neutron and there is also an electron. It is a simple diagram of a bar magnet, in where a proton is on one side, the neutron is on the middle, and the electron is on the other side, [(+) (0) (-)].

And because of the Big Bang the three main parts or the three main solid pieces of the atom or the magnet are separated from the center or separated from the nucleus.

The proton and the remaining neutron stay at the center or at the nucleus, but the electron takes position in the magnetic field or in the so-called electron or magnetic ring, [(+) (0)] (-).

Actually, the magnetic field or the magnetic ring is really a neutral or is an electrically neutral ring or field.

When the electrons are separated from the center or from the nucleus of the atom, the electron occupies the neutral ring making or turning the neutral ring into an electron ring or electron field or shield.

The electron transforms or charges the magnetic ring or shield into an electron ring or into a negatively charged ring,

To simply find the exact number of electrons or the exact number of negatively charged pure matter particles orbiting or going around the nucleus of an atom, one must count by two for the very first electron or for the very first magnetic ring and whatever number of electrons or whatever number of negatively charged pure matter particles remains goes to the second electron or to the second magnetic ring.

Helium, for example, has two electrons or has two negatively charged pure matter particles on the very first magnetic or the very first electron ring or the very first magnetic or electron orbit.

In reality, the Element Helium is an element made up or composed of two magnets united at the corners.

And being made up or being composed of two magnets, the Element Helium has two protons, two neutrons and two electrons, $[(+)(0)(+)(-)(0)(-)]$.

Turning to the third element or the three whole or real numbers, count two electrons for the first electron or the first magnetic ring and count one electron for the second electron or the second magnetic ring.

And finally, turning to the fourth element or to the fourth atom, count two for the first electron ring, and there is a remainder of two for the second electron ring.

The fourth element is really four magnets united at a corner or united in links.

The same example or physical pattern as above follows on until the next neutral element or the next neutral atom, which is a gas.

Pure matter as gas or an element as gas is not electrically charged.

The next neutral element, which is Neon or element number ten, follows the very same and simple example or simple physical pattern of the first elements before Neon: count two for the first electron ring or the first magnetic ring, and count eight for the second electron ring or for the second magnetic ring.

The elements following Neon or following element or atom number ten follow the physical pattern of Helium, two electrons, and Neon, eight electrons.

The pattern for Neon is two for the first electron ring, eight for the second electron ring and whatever number remains goes to the third electron ring or goes to the third magnetic ring.

Element number eleven is counted two small cubes for the first electron ring, counted eight for the third electron ring or the third magnetic ring.

Element or atom number twelve is counted two for the first electron ring, counted eight for the second electron ring and counted two for the third electron ring or the third magnetic ring.

The very same physical example or the very same and simple physical pattern as above follows on until the next neutral element, and the very same and simple physical procedure as Helium and Neon must also follow.

That is to simply say, the next neutral element determines the number of electrons in electron rings or in magnetic rings.

The next physical break or the next physical division of the Periodic Table of the Elements, for example, determines the exact number of electrons that are going to the electron rings or to the magnetic rings.

It is already pre-determined even before matter or the original atom enters the universe or the vacuum of space.

As it can also be simply seen, the atom or the divisible atom is not only a simple microscopic diagram or a simple microscopic blue print of a bar magnet or a simple microscopic diagram of an electric motor or generator.

But also the divisible atom is a very simple microscopic diagram of the physical universe and of physical Existence or of physical reality or of physical creation and even the conscious mind and brain.

The divisible and the physical atom is actually a simple magnet inside another simple magnet, but instead of the two simple magnets physically attracting or physically uniting as two opposite charged magnets, the two atomic or the two microscopic magnets are kept physically apart by the magnetic sphere or by the magnetic field or shield or simply by a magnetic shell.

In a way, the magnetic sphere or the magnetic field of the divisible atom acts as a simple protective shield or acts as a simple protective sphere or act as a simple protective shell.

The simple magnetic sphere or the simple magnetic shell, therefore, keeps the divisible atom from physically collapsing or from further physical expansion or from further physical separation from the nucleus or from the center of the divisible atom.

That is the very simple reason that the element or that pure matter does not become physically much smaller or microscopic as the divisible atom.

In other words, the simple magnetic field or the simple magnetic sphere or the simple electron ring of the divisible atom or the element keeps matter from further physical expanding or from further physical collapsing or from physically becoming microscopic or from even physically imploding.

However, this does not mean that the magnetic sphere or that the electron ring cannot be acted upon from outside.

In contrast, interacting with the divisible and the physical atom from the outside creates mixtures of new elements or two or more elements united, thus making other elements or what is simply called in Modern Chemistry isotopes, compounds or mixtures.

This, however, does not mean that there will ever be more than 118 elements.

The Small which makes the Great

[(+) (0) (-)]

Existence is infinite or without size. Existence runs or she expands toward all sides at the very same time and at the very same time existence also compresses toward all sides, that way also adding to her very self an infinite weight and an infinite size.

That is to say, existence adds to herself, she multiplies herself, she divides herself and existence also subtracts her very self or from her very self to be able to become forever for more and even for new.

Existence forever existed and existence forever will exist. Existence has no time even though time existence is.

Existence is also composed of three principal parts which all add to one and that one is the grandiose part which makes the difference in all of existence.

Existence consists of matter, that which is physical. Existence also consists of space or of vacuum, that which is lack of something or of matter.

And existence also consists of movement or of that which many call space-time, which is a movement as if in waves in space.

Thus in truth, existence is a simple magnet!

In other words, existence really is composed of +0-; in where (+) is equal to positive; in where (0) is equal to neutral; and in where (-) is equal to negative or the lack of.

In the scale of colors would be white, grey and black; in where white is light or matter; grey is neutral or space-time; and black is darkness or lack.

Now then, just as existence truly is composed of three parts and practically of parts opposite to the other parts, +0-, thus one also truly is composed of those very same opposite parts but in knowledge because one is knowledge.

And one can truly be positive knowledge or be neutral knowledge or even be negative knowledge.

But the greater the positive knowledge one has, less the neutral knowledge and less the negative knowledge or less the lack of negative knowledge.

In other words, the greater the knowledge of reality that one has, the greater the reality of one and less is the reality not known or less is emptiness or darkness.

Thus in truth, the more the knowledge of one, the more the abundance or the more is the life of one or more life makes sense to one.

Interestingly, that existence functions the same as one but existence knows it not but existence will never stop from being or will never stop from existing for all the eternity of eternity because she truly makes herself for much more!

That is in truth, even though all of existence is one, because of one existence truly is and for all eternity infinite!

And every time that existence is increased by one or is increased by more than one, the point of neutrality or the neutral is less as the same as the negative or the unknown point is less.

But making the things less does not make existence greater, but making the things or existence greater makes existence greater!

And just as existence makes the things greater for existence to be greater herself, oneself can be greater by recognizing that life is more or that in life herself there is more and that there is with all peace, the neutral, and that here is with all knowledge, one or more than one; and also that there is with all gladness and with all joy and that also there is with all abundance of life, life renewed…

Thus, through knowledge one renews for much more and because of one renewing for much more, one truly can continue for all of eternity as the very eternity…

The Element

[(+) (0) (-)]

The universe is a neutral point or is a point zero which is converted into a positive point or into a one or more when light or matter enters into the universe.

The universe also is converted into a negative point when the grand majority of light turns off or the majority of matter no longer has energy...

Before light or matter enters into the universe the weight of the universe is cero and when light or matter enters into the universe thus the universe takes on weight.

Light or matter is composed of 118 Elements and the weight of each element is two times its atomic number, more like its positive number.

In the case of element number one, for example, its weight is of two and in the case of element number two its weight is of four and in the case of element number 118 its weight is of 236...

Curiously, that the totality of the 118 elements adds to one and that the totality of their weight adds to 2.

That is to say, that if we added from one to 118 thus the sum would be 7,021.

And if we added that sum thus it would be 10 and if we add that last sum thus it would be one.

That is to say, 118 is equal to one...

And if we added from two to 236 the sum would be 14, 042.

And if we added that sum thus it would be 11 and if we added that last sum thus it would be 2.

That is to say, 236 are equal to two...

Thus, light or matter enters into the vacuum of space or into the universe as one or as a unit which is composed of 118 pieces or elements and the weight of the element is two times the atomic number of the element.

Thus in truth, number one itself is composed of 100 percent plus 18!

That is to say, that the number one or even oneself is equal to 118 percent!

Also light or matter could enter into the vacuum of space or into the universe with only or as one element with its weight of two but that element would be able to convert into the other elements, even to the element 118 and its double weight of 236...

Element number one, for example, enters into the vacuum of space with its weight of two and it has 117 other possibilities of converting into the other 117 elements.

That is to say, element one is composed of 118 parts or pieces or the 118 percent and element number has 117 possibilities of converting into the other 117 elements according to the weight which element number one maintains.

In the same manner, element number two with its weight of four has 116 possibilities of converting into the other 116 elements or until the element number 118 with its weight of 236...

When light or an element enters into the vacuum of space, light or the element enters as if it were a piece of magnet or as if it were a bar magnet.

In the vacuum of space or in the universe the magnet or light or the element is super compressed not only until it takes the form of a sphere or round but also light or the element or the magnet is super compressed until it gets to a very high level of temperature.

And when the temperature gets to its highest level thus light or the magnet or the element super explodes causing the light or the magnet or the element to divide into two parts and the part with less weight, such as the negative part, takes position in the magnetic field and that magnetic field now is a negative field...

The superior or the positive or the heaviest part of light or of the magnet or of the element takes position in the center or in the nucleus.

Thus, now we have the negative part of the magnet going around the positive part when before they were united and the center or the nucleus was neutral...

And even though the element lost weight because of the explosion or because of bursting, the element continues the one for two.

That is to say, its weight continues of two although the element one now is 0.999 and its weight is double, 1.998...

Thus, now element number one was reduced to about 0.999 with its new weight of 1.998 but to element number one also remains a neutron or even more than one or a neutral part which can be converted or can be transformed into a positive part and that way not only adding the number of the element but also adding its weight even though it will have only one negative particle going around the center or the nucleus or the positive side...

But if one or light or the magnet or the element does not convert into the next number or into the next element thus it loses its energy and will only be a piece of dead matter in the vacuum of space and it will be removed one day by the black holes...

Thus, light or the element is the same as a negative particle, is the same as a neutral particle and is the same as a positive particle.

In a way, one is equal to a negative portion plus a neutral portion plus a positive portion which totality is of 0.999 after entering the vacuum of space.

But in the vacuum of space light or the element is positive even though the vacuum of space is neutral but reacts as if negative because of the vacuum.

And when the element increases its positive part by converting the neutral part into a positive part, the element cannot attract the negative part because of the vacuum of space because now the negative part becomes as if more or its weight increases because of the weight that it receives indirectly from the vacuum of space.

And if there were not positive attracting the negative thus the negative would expand through the vacuum of space and it would stop being negative and it would be dead matter...

Now, an element, in this case a star, which number is high as the same as its weight thus lasts or remains longer in the vacuum of space or in the universe.

But the element or the star becomes heavier while the energy or matter to continue on lasts or it begins to transform from positive to neutral and once neutral, the element or the star practically becomes negative when its excess or super weight attracts the electrons toward the center and thus causing an implosion in where the element or the star becomes a nova or a new star but without light or without energy and that way causing an enormous hole in the vacuum of space when before the element or the star occupied the vacuum of space as an element or as a star...

In other words, matter or the star in the vacuum of space changes from positive to neutral and then from neutral to negative.

Thus, [+ 0 -], in where the negative is a black hole or is a super vacuum cleaner in the vacuum of space or in the universe or in matter or the element.

This black hole practically eats or sucks all the matter around it to take all matter out from the vacuum of space or from the universe to make new space for new matter or for another beginning...

But as long as there are black holes in the vacuum of space or in the universe, thus the vacuum of space or the universe is negative and as long as the vacuum of space or the universe continues as or is negative thus it keeps being for something and not for neutral and from neutral or from zero to positive...

Thus, so that the universe becomes neutral or to zero and from neutral or from zero to positive thus all the black holes or super space vacuums must stop from functioning and once the black holes or the super space vacuums stop from functioning for lack of matter or for lack of energy thus the vacuum of space or the universe will return to neutral or to zero...

Now then, this new vacuum in space or in the universe is neutral and has zero energy but even so attracts new matter or new stars or attracts the new or the next beginning which is outside of the universe or outside the vacuum of space...

And once new matter or new stars enter into the vacuum of space or into the universe thus the vacuum or the universe will be positive or one or more...

Thus, the cycle or the model or the rhythm of (+ 0 -) continues until the end of all the times...

Summing up the atom, the atom as also is the universe is finite or ends but existence is infinite or without size and has no end.

The atom or the universe ends or stops running but existence runs or she expands toward all sides at the very same time and speed and at the very same time and speed existence also compresses toward all sides, that way also adding to her very self an infinite weight as also an infinite size.

That is to say, existence adds to herself, she multiplies herself, she divides herself and existence also subtracts her very self or from her very self to be able to become forever for more and even for new or renewed but forever in perfect abundance and as if forever as if she was always new or renewed.

The universe begins as matter enters the vacuum of space, but existence forever existed and existence forever will exist. Existence has no time even though time existence is.

Existence is also composed of three principal or main parts which all add to one and that one is the grandiose part which makes the difference in all of existence.

Existence consists of matter, that which is physical. Existence also consists of space or of vacuum, that which is lack of something or lack of matter, energy or light.

And existence also consists of movement or of that which many call space-time, which is a movement as if in waves in space, more like a magnetic field.

Thus in truth, existence really is a simple magnet!

In other words, existence really is composed of [(+) (0) (-)]; in where (+) is equal to positive; in where (0) is equal to neutral; and in where (-) is equal to negative or the lack of.

In the scale of colors would be white, grey and black; in where white is light or matter; grey is neutral or space-time; and black is darkness or lack.

Now then, just as existence truly is composed of three parts and practically of parts opposite to the other parts, [(+) (0) (-)], thus one also truly is composed of those very same opposite parts but in knowledge because one is knowledge.

And one can truly be positive or real knowledge or be neutral knowledge or even be negative knowledge, just as false belief.

But the greater the positive knowledge one has, less the neutral knowledge and less the negative knowledge or less the lack of negative knowledge.

In other words, the greater the knowledge of reality that one has, the greater the reality of one and less is the reality not known or less is emptiness or darkness.

Thus in truth, the more the real knowledge of one, the more the abundance or the more is the life of one or more life makes sense to one.

Interestingly, that existence functions the same as one but existence knows it not but existence will never stop from being or will never stop from existing for all the eternity of eternity because she truly makes herself for much more!

That is in truth, even though all of existence is one, because of one existence truly is and for all eternity infinite!

And every time that existence is increased by one or is increased by more than one, the point of neutrality or the neutral is less as the same as the negative or the unknown point is less.

But making the things less does not make existence greater, but making the things or existence greater makes existence greater!

And just as existence makes the things greater for existence to be greater herself, oneself can be greater by recognizing that life is more or that in life herself there is more and that there is with all peace, the neutral, and that here is with all knowledge, one or more than one; and also that there is with all gladness and with all joy and that also there is with all abundance of life, life renewed…

Thus, through real knowledge one renews for much more and because of one renewing for much more, one truly can continue for all of eternity as the eternity herself…

Chapter Five

The Human Brain

Depart from physical reality and physical reality will be seen even though one really enters into desolation for lack of knowledge of that higher or taller state!

The human brain, for example, will complete or will fill that which the naked human eye actually or really cannot see, but first "that" must be looked at and seen or even presented or confirmed or reconfirmed first!

Draw two lines together or intersecting one way or another and the human brain will add a third line, trying, therefore, to make something complete or something meaningful of the first two lines!

Draw a half circle and the human brain will try to make something of that half circle or add to that half circle, trying to make that circle full or complete and understandable or useful and thus enjoyable.

Too often, therefore, a completed picture or a completed diagram or a drawing can only be seen or can only be fully understood and fully enjoyed when there is a bit of physical motion of the naked human eye or a physical repositioning of the interested observer or the interested learner.

And even though the physical motion is not from the picture or from the diagram or from the drawing, that physical motion or point of view or the physical repositioning of the interested observer or the interested learner truly helps to complete, to simply see or to fully and simply understand and thus fully enjoy the complete picture or the complete diagram or the complete drawing!

It is as if the physical motion or the point of view of the naked human eye or the personal interest or the physical repositioning of the interested observer or the interested learner fully completes or complements the picture or completes or complements the diagram or drawing even though the diagram or the picture or drawing truly was already complete or was already perfect.

In other words, a bit of physical motion or change or improvement of point of view or the physical repositioning of the naked human eye helps to fully understand and thus enjoy fullness or really enjoy actual reality!

That is to simply say, the personal point of view or personal interest or the physical repositioning of the observer helps to complete or to make perfect existence and thus enjoy existence in and through enjoyment or gladness and joy, both responsible for rebirth or life anew!

Explaining the true origin of the universe or of creation or of existence or of reality or even the true origin of time, and its true physical inner and its true physical outer workings, requires both a theory or idea or belief or some personal interest or point of view and some solid or some physical scientific facts or solid bits and pieces.

And it also really or actually requires describing something solid or something physical and something that the human being cannot really or actually see in and out in its entirety (with some of the solid pieces or some solid scientific facts or solid bits and pieces that the conscious or that the thinking human being already has or simply thinks or simply believes that he may have but sadly has not).

In other words, it takes some physical reality or some solid bits or some solid pieces of physical existence or actual reality or physical creation to further explain physical reality or physical existence or physical creation and thus explain the physical origin of the entire and the physical universe and physical or actual creation!

It takes solid or physical facts or solid bits and pieces with some personal interest or good or better point of view to truthfully explain physical existence or actual reality or creation and its true inner and its true outer workings!

The personal interest or point of view is the physical and the actual motion that will physically put together the solid or the physical facts or the solid bits and pieces and, therefore, complete the physical universe or complete physical existence or complete or finish or confirm actual reality or complete physical creation through acknowledgement or reconfirmation!

The human being, for example, cannot see his own physical face or see his complete or true physical face in front of him unless there is a solid or a physical reflection or a real or actual physical sensation of some kind!

But when he finally gets to see a solid or a physical refection or feels his solid face or complete or true physical face, the human being can physically see or can physically sense or can even physically feel the exact physical opposite of what he is really or actually feeling, really or actually sensing or even really or actually looking at!

And it is really or actually easier to theorize or it is easier to fantasize or than it is to truthfully or to completely believe or to truly or to completely see or to fully and to understand physical reality as physical reality or as physical existence really or actually is and how physical reality or how physical existence really or actually works.

The true or physical reality, or simply the naked truth, and the sad irony of the physical and the whole matter is that the conscious human being will never be able to see his true or real and physical face or complete face as his true or real and physical face or complete or actual face really is in physical reality or in physical existence no matter how good and simple the solid or the physical refection or how great the real and the physical sensation!

It is not because of constant physical change or constant physical motion or that the conscious human being physically grows old, gray and blind that he cannot thus really see his true face or complete or actual face, but because the conscious human being, as of yet, cannot really see or cannot fully understand physical reality or cannot understand physical existence as physical reality or as physical existence really is!

And not only that, but enjoying physical reality or really enjoying physical existence is also a major part of physical reality or physical existence, no matter how physically limited the actual or the present reality or the actual or the present and physical existence!

A complete, a perfect or a full face is a face with a great or joyful smile, but once that great or joyful smile is gone, so is the complete, the perfect or the full face!

Therefore, to have a complete, a perfect or a full face, the conscious living being must also and simply have a complete, a perfect or a full mind!

A complete, a perfect or a full mind is a mind with complete, with perfect or with full or with true knowledge of reality or actual existence or even creation!

True or naked knowledge of reality or actual existence or creation is that knowledge that not only brings forth more true or more naked knowledge, but true or naked knowledge also expands the human mind through complete or through full gladness and joy or through total or through complete or perfect enjoyment or happiness!

In other words, naked or complete knowledge of actual existence or of physical reality or of actual creation really opens or really gives more visions to the naked human eyes so that the conscious human being or the conscious living being can really or can actually see the naked, true or complete physical reality or actual creation that already was or that really or actually is in front of the conscious living being.

The function, the purpose or the job of the natural human mind or the function or the job of the natural human receiver, or simply call it the animal instinct or the animal mind, is to see in solid parts, in solid bits or in solid pieces, and not see whole or complete!

The animal mind cannot see in full or cannot see in fullness. The animal mind or the animal instinct cannot really see or cannot really understand in fullness.

The human being, for example, cannot see the whole picture or cannot see the whole diagram or cannot see whole drawing unless the human being first sees physical parts or first sees physical or solid pieces of that whole or that complete picture, or complete diagram or drawing, or the whole picture or diagram or drawing is in physical motion or the whole picture or diagram or drawing smells and then the human being reacts or takes notice.

It is also an animal instinct to see in physical or in a solid part or in solid pieces and to physically react to the physical motion or react to the physical sound of the physical or actual motion or react to the new point of view.

The animal or the natural mind or the natural instinct, for example, cannot really or cannot actually see a physical whole or really cannot see a complete tree.

That is to say, the animal mind or the animal instinct does not allow the animal to see a whole or a complete tree.

An animal mind can only see solid parts of that physical tree; and even though that animal mind can see different solid parts of a physical tree, that animal mind cannot really see the whole or cannot see the physical and the actual tree!

In the same manner, animals or insects, no matter how smart or even well trained, can only see light and shades.

Again, the animal mind or the animal instinct or really point of view cannot really see the physical sun, cannot see the physical stars or even cannot see the physical moon, but can only see the physical light and the physical shade.

It is through the shade and the flickering or the bit of physical motion that the human being through his animal or natural mind or through his point of view can see physical reality or can really see physical existence as physical reality or as physical existence or even physical creation really is!

The solid or the physical part through some personal interest or through some physical motion or point of view really explains the solid or the physical whole.

The solid or the physical whole really cannot or actually cannot be fully or completely seen or cannot be completely understood unless there is some physical motion or there is something physically doing the addition, the division or the break-up!

To describe or to analyze something physical or physically, the scientist, the researcher or the writer, or even a police detective, therefore has to either physically take it apart or to put it physically together from the solid bits or from the solid pieces the scientist, the researcher or the writer, or even a police detective, already has or the solid bits or solid pieces he sees or understands!

From the missing solid bit or from the missing solid piece or from the missing solid pieces or from the missing solid bits the scientist, the researcher or the writer, or even a police detective, theorizes or makes an educated guess or develops a new point of view, thus making physical reality or making the solid fact or the naked truth very difficult to see or very difficult to really understand and very difficult to prove and thus enjoy.

The theory, the educated guess or the personal interest or the new point of view is, therefore, the glue to glue together the solid bits or the solid pieces or the physical reality or the solid facts or the naked truth that the scientist, that the researcher or the writer, or even a police detective, already has, but too often the glue or the theory or the reasoning or the educated guess or point of view is more than the reality, and ironically makes but more puzzles or more solid pieces than those that the scientist, the researcher or the writer, or even a police detective, already has or actually or really started with!

What a sticky mess!

And so most writers and most researchers as most scientists, sadly but true, become more interested in the good sounding or the good smelling glue, therefore filling countless books with half-truths or with false knowledge having nothing to do with the physical reality, the naked fact or the solid pieces that the writers, the scientists or even the researchers actually started with!

What sad irony and what a waste of valuable time and what a waste of precious physical resources!

Confusion is not knowledge, but confusion is the lack or the misunderstanding or the misuse of knowledge!

That is to say, true or real knowledge is physical and actual reality; and the lack of true or real or naked knowledge is not physical or is not actual reality.

In other words, complete or full or perfect knowledge is complete or is full and physical reality. And complete or full reality is but complete or full enjoyment; and full enjoyment truly brings an expansion or an abundance of conscious mind!

And that is the actual reality or the naked truth!

In other words, real or true knowledge, or naked truth, really brings forth a useful expansion or brings forth a useful abundance of conscious mind or brings forth an expansion or an abundance of conscious thought.

An expansion or an abundance of conscious mind or conscious thought brings forth more real, truer or more very useful naked knowledge. But that is simply not happening!

Even well trained or highly experienced scientists, researchers or professional writers are still very prone or very sensitive to their own point of view, personal prejudices, to their complex personal problems, to their cultural or even to their religious beliefs or to their lack of solid scientific or the lack of solid information or the lack of solid or naked knowledge or even the misunderstanding or the misuse of true or naked or physical knowledge!

For example, which of the two following questions has a real or has a possible or complete answer?

Which of the two following questions really or truthfully has a solid or has a physical answer?

Which of the two following questions really has a solid or has a physical foundation?

"Why is the Universe round or sphere-like?" Or "What makes the Universe round or sphere-like?"

The "Why?" or the very first question is a very childish, is a very foolish or is a meaningless and a useless question!

The "Why?" question really limits point of view or really limits way of thinking. The very childish, the very foolish or the meaningless and useless question is asked as if no real, no actual or no true physical or complete answer is really expected from the person asking the very childish, the very foolish or the useless and thus meaningless and incomplete question!

At the very same time, that very childish, that very foolish or that useless and thus meaningless question keeps the person from simply seeking an actual or a real or a physical or complete answer simply because that very childish, that very foolish or that useless and thus that meaningless question sounds very complicated, sounds very unreal or sounds simply unanswerable!

And if the very childish, the very foolish or the useless and thus meaningless and unanswerable question were actually or really asked to another person, that other person too would be dead stuck and would be dumb founded, and would not even attempt or would not even bother or simply would not be personally interested enough to even give an answer, any answer for the answer too would be very childish, very foolish, totally meaningless, and without any real understanding and thus without any real or without any true or without any meaningful enjoyment!

The "Why?" or the very first question is like an adjustable pipe wrench or like a simple tool deliberately thrown into the gear box that is the human brain!

In other words, the "Why?" question, if it is actually or really a question, is a very silly joke that no one really understands and because no one really understands the silly joke, no one can really appreciate the silly joke and really cannot enjoy the silly and meaningless joke!

Remarkably, however, that very childish, that very foolish or that useless and thus meaningless and unanswerable question is asked by most scientists, by most writers and by most professionals than it is really asked by a very funny or a very silly or a very humorous child or simply asked by very curious and very playful children!

On the other hand, "What makes the Universe round or sphere-like?" has an answer, a very simple and a very satisfying, understandable and thus a very enjoyable and a real and a physical and complete answer!

In fact, part of the satisfying and understandable and real and physical answer is already in the answerable question because the question is a real or is a complete question!

The answerable and the real question is already telling us that something is, therefore, really making the Universe round or sphere-like!

It is a very simple but true question that although will lead to many more simple and many more real or true questions, the simple questions will also give one final, one simple and one truthful or one complete and one very satisfying answer, an answer that can be fully or can be completely understood and thus simply enjoyed!

However, it does not really take too much creativity, too much imagination or too much inventiveness or too much glue or even too much of a theory or educated guess or even point of view to simply see, to simply understand or to simply discover and thus simply enjoy the naked truth or the simple physical or solid reality or physical creation that is already in front of mankind or already was in front of human kind since the very beginning of conscious thought or point of view!

The naked truth or physical or solid reality, as it was already and stated above, and its simple inner workings and its simple outer workings, is simply in front of or before mankind.

In fact, mankind walks right through physical or complete reality every day; but because mankind already is a physical part of what mankind is very hard trying to explain, to see or to complete or to simply understand, mankind simply cannot see the whole or cannot see the complete picture, especially if mankind is the one doing the physical motion or doing the constant but simple change or continuation!

And because mankind usually and really questions or usually and really asks for the wrong reason or reasons, mankind simply cannot see in the simple light or even sense the simple light and cannot walk in the simple light.

Mankind becomes more of what he does not really want to be with knowledge or the solid pieces he thinks that he may have, but ironically he really has not!

And what is the naked truth or the physical or the solid reality or the hidden physical truth, and its simple inner workings and its simple outer workings, that already is before or in front of mankind?

The simple reality or the complete or naked truth is that mankind does not let his very own brain do its wondrous and enjoyable job!

The human brain or human conscious thought is more than a simple copy or more than a simple duplicate of physical existence or physical reality or even creation. Mankind uses his conscious mind, a very small portion of the human brain, to simply misinform or to even poison and to cripple on purpose the human brain!

One way or another, mankind on purpose limits his point of view or really covers the simple reality or the simple naked truth or filters out the proper physical information or filters out the proper or the true knowledge of reality or actual existence or creation that will let the human brain to simply balance itself with true or with complete physical reality or with true or with complete physical existence or actual creation, which simply is in constant but simple change or continuation!

The human being purposely and simply incases or imprisons his human brain by purposely incasing or imprisoning his human brain by feeding the human brain false knowledge of reality or actual existence or creation or by feeding the human brain half-truths or by feeding the human brain poison with false beliefs or ideas!

In other words, if the human brain does not have the proper physical or does not have the correct information or the proper or the correct recall or the proper or the correct memory of actual reality or physical reality or actual existence or creation, there is but total physical chaos. The human brain is already physically dead before the due time!

For example, when a physical part of the human body no longer reacts to physical pain or to physical sensation or the physical body part is motionless, not giving proper or correct physical information or proper or correct physical feed back to the human brain, the human brain will physically shut off that misinforming part of the human body!

To the human brain, therefore, correct or true physical information or complete or truth knowledge of actual reality or physical existence or creation is more important than is a malfunctioning or misinforming body part or even a misinforming human mind!

And that is the simple and the naked truth or the physical reality thus far!

What, then, does this work have to do with the true physical origin of the entire Universe or true reality or true creation, and its true inner and its true outer workings, one might simply ask?

Ironically, the most difficult part of this work, or this simple glue, has being the introduction herself. Physical lack of true or complete knowledge or lack of knowledge of actual reality or of creation is not the only thing that keeps mankind or the human race physically and mentally from moving progressively or successfully forward in time or in history, but also that which mankind thinks or believes that he simply has but actually has not!

The sad reality or the sad fact or the sad and naked truth is that the human race has followed the very same simple line or the very simple pattern of thinking or the very same reasoning or point of view from as far back as 6,000 to 12,000 years or more!

The so-called human or the so-called scientific knowledge or point of view simply does not really add to physical much.

The human race or mankind really or actually is still living in the dead past, with the very simple physical exception that his modern brain shuts off parts of the his human body simply because of lack or very poor information, or simply because of lack or very poor solid facts or the lack of physical information or poor point of view of actual reality, of physical existence or the simple lack of information of simple human surroundings, natural or unnatural!

And knowing the true physical or the true origin of the entire Universe or of creation or of all physical existence or even knowing the true origin of physical time, and its simple inner workings and its simple outer workings, will not really or simply will not add to physical much or to anything unless mankind simply sees or simply knows or simply understands completely or in full what actual or physical reality or actual or physical existence or creation is all about.

And once the conscious living being really or actually sees reality or creation as reality or as creation really is, reality or creation is no longer the same.

Once again, the very simple reality or the complete truth or the naked and the very simple truth, lets call it that, is in front of mankind.

In fact, mankind walks right through actual or physical reality every day. But because mankind simply and usually questions, senses or feels, or even fears, for the wrong reason or reasons, thus mankind simply cannot really see in the simple light.

Of all the living creatures, man does not only break all the physical laws of physical nature, but also man is the worst interpreter of all the physical laws of physical nature!

In fact, if man, from the very beginning of consciousness, really understood or really recognized the simple naked or the simple physical laws of physical nature, none would have being broken, and man would be so much ahead in conscious mind and in physical body!

Sadly, however, mankind simply becomes more that complicated physical law breaker or that confused physical creature or that useless or that complicated animal or that complicated or useless physical thing that mankind so much try so hard not to simply be.

Sadly, mankind is simply becoming an unnatural animal or a physical creature of solid but useless stone or useless mud or clay!

Mankind is becoming a cripple and useless clay animal that cannot even and simply drag itself into the simple light!

As it can be simply seen, the very simple and the actual physical reality or simply the simple naked truth is that mankind, once again, simply does not let the human brain to simply function!

That is, the human being does not let the human brain do the brain's wondrous and very simple and enjoyable job!

The human brain simply is more than a simple physical duplicate of actual existence or actual or physical reality or even creation herself!

Mankind uses his simple conscious mind, which is a very small portion of the human brain, to deliberately misinform or to deliberately poison the human brain!

The human brain, being a simple duplicate of physical existence or creation, has true or has complete or has full knowledge of all physical existence or all physical reality!

One way or another, mankind simply covers the simple and the naked truth or simply filters out the proper physical information that will let the human brain to simply and to actually balance itself with actual or with true reality or with actual or with true existence or actual creation, which is in constant physical change or in constant physical motion!

In other words, if the human brain simply does not have the proper or the correct physical information or the proper recall of that physical information or the correct point of view, mankind has once again but total or complete chaos!

Actually, there simply cannot really be such a thing as total or complete or full chaos; but almost the exact opposite.

It is the simple lack of knowledge or confirmation or the simple lack of acknowledgement or reconfirmation or too much misinformation that causes chaos or that really causes confusion.

Confusion is simply the lack of understanding, limited point of view or chaos is simply the lack of physical matter or simply too much matter or information pressed together, which is simply knowledge or confirmation, which is not responded to or which is not acted upon.

Confusion is facing a simple mud or brick wall and not being able physically to simply step back a few steps and not being able to simply understand or not being able to simply and physically see beyond the physical mud or the physical brick wall!

That is to simply say, total, full or complete chaos is too much lack of physical or too much lack of solid information or simply too much misunderstanding or too much confusion simply happening at once or too much lack or too much misunderstanding or too much confusion happening one after the other.

Total, full or complete chaos is simply useless blanks or simple cracks where there should be none.

In short, the simple reality is that the human race does not let the human brain to simply complete physical existence or to simply complete physical reality or to simply complete or continue simple creation!

The human brain is simply a reality or existence maker, and the human or mankind does not let the human brain do so. And that is really the simple but the naked or the complete truth!

When the human brain, even a simple animal brain, does not get correct or useable physical feedback through the different senses, the human brain will shut off or will shut down the misinforming senses!

The human brain will even go as far as shutting off physical parts or even shutting off the whole human conscious mind or the whole physical body or both to simply stay in simple physical tune or in simple physical balance with the physical universe or with physical existence or with physical or actual reality or with creation herself!

Sadly, that simple shut down of different or many senses, body parts or simple shut down of the human conscious mind or conscious thought goes on or continues on into many future generations!

Some people are born blind in one eye or both, and parts of their conscious minds, rational or irrational, as well as physical parts from their bodies, do not function or do not work simply because of the lack of brain information or the simple misunderstanding or the simple confusion or limited point view or chaos during their own generations or their past generations!

Sadly, however, modern or even future medicine simply will not kick start that which the human brain simply has shut off!

In other words, human science or human medicine or human technology simply and really cannot deceive or really cannot trick the human brain because human science cannot understand physical existence or physical reality or creation!

But once human science or the conscious living being understands physical existence or actual reality or creation, there will be no need to try to deceive, no need to try to lie or no need to try to trick the human brain into more physical freedom and into more mental or conscious freedom or into more conscious thought.

Many thousands of years ago, the human brain felt safe enough with the very simple physical information, point of view or with the very simple physical knowledge of physical reality or actual existence or actual creation at hand and gave liberty or released the human conscious mind or conscious thought.

Sadly, however, that simple release or that simple freedom of the human conscious mind or conscious thought created the thing called the Pandora Box or more like the Pandora Jar because the interpreter or translator misinterpreted or mistranslated the box for a jar!

That simple release or that simple freedom of the human conscious mind or conscious thought by the human brain simply created but total or a complete chaos, a total or a complete chaos that still goes on today and every day, and will still go on for some time!

That is, until mankind does something about the human conscious chaos, limited point of view or the human conscious confusion or the human conscious condition that mankind simply bestowed upon himself and bestowed upon many future generations!

The human conscious mind

Knowledge is confirmation and acknowledgement is reconfirmation, something which the human brain is really waiting for!

The simple and the single purpose of the human conscious mind or reasoning or human point of view or conscious thought, therefore, is to simply gather or is to simply bring about more and complete or true physical knowledge of reality or actual existence or even physical creation.

And that is the very simple reality or the very simple physical and the very simple naked truth!

That is to say, true or naked knowledge of physical reality or actual existence or physical creation brings truer or more naked knowledge of physical reality or actual existence or physical creation, especially when there is a very simple personal or a very simple mental or conscious interest or point of view and thus allowing for more personal enjoyment of true or complete or naked knowledge!

Instead of giving mankind or the human race simply more human conscious mind or more physical room for conscious reasoning or better point of view, the human brain has simply taken away some parts or taken away some control of the human conscious mind or conscious thought simply because the human brain feels very threatened or feels very insecure by the very freedom or the very liberty of the human conscious mind or the human conscious thought or point of view of the conscious mind!

So instead of physically walking erect with the two legs straight in solid ground and with the conscious mind high in the clean air, mankind is physically bending to soon crawl in mud on four legs once again.

That is to simply say, if mankind still has moving body parts; but if not, mankind can physically drag like a simple worm in the simple bed of dust or bed of mud or bed of disgorged matter or bed of disgorged knowledge mankind is also and purposely creating.

No conscious living being willingly eats vomit or willingly eats digested matter. The same rule thus follows for the human brain.

Not until true or real knowledge of physical existence or actual reality or creation personally hits the conscious living being, knowledge of physical existence or actual reality or the naked truth will not be appreciated or will not be fruitful and thus simply will not be enjoyed and will not be rewarded with further expansion of conscious mind.

Knowledge of physical existence or actual reality or creation is simply for the simple benefit of every conscious living being.

Knowledge of physical existence or knowledge of actual reality or of physical creation is like a physical apple. Every conscious living being simply recognizes the very simple physical apple or a simple fruit, but every conscious living being appreciates more the very simple physical apple or the simple fruit when the very simple physical apple or the simple fruit has being personally tasted and has physically expanded his stomach.

The very simple physical expansion of the stomach, therefore, made the very simple physical apple sweeter!

In other words, the very simple physical expansion of the human conscious mind or a better point of view through conscious thought really makes physical existence or creation but much sweeter!

So, therefore, the conscious living being has to personally get involved with true or with complete physical knowledge of existence or of actual reality or of creation even though that complete or that true physical knowledge of existence or of actual reality or of creation came from other conscious living beings.

How long does the human race, therefore, has to wait before the human brain feels physically safe once again, not only to release those parts already taken away from the human conscious mind, but feel physically safe enough to release new and more mind parts or more conscious mind freedom or allows for new point of view?

Does the human race have to wait another three to four thousand years or even much more to feel physically save or secure?

Unfortunately, the human individual does not live that long; and even if the human individual did live that long, what a waste of time!

But, fortunately, the human race does have something very unique and very simple which can penetrate mountains, which can penetrate borders or even can penetrate time.

It can even travel faster than the speed of light without the friction. It is simply called conscious thought or conscious reason or reasoning or even acknowledgement!

But conscious thought or conscious reason, as is the human conscious mind, is also a very heavy burden to the human brain!

Instead of that conscious thought or that conscious mind simply penetrating through solid rocks or simply penetrating through solid barriers or simply filling the missing parts or the missing pieces of knowledge or filling the gaps of information of physical reality faster than the speed of light, the human race simply becomes very heavy burdened or too heavy and too stiffened and simply cannot physically move and thus simply cannot see or simply cannot walk in the simple light and simply but enjoy!

That is to say, the human race uses a lot of energy and uses a lot of valuable resources or uses simple knowledge or simple information to simply keep but still!

In fact, the human race lives with old or obsolete or useless knowledge or useless physical information of physical reality or actual existence!

Instead of truly increasing the old knowledge of physical reality or physical creation with the new but true knowledge of physical reality or physical creation, no real room is really or is actually left for the new or for any real improvement or advancement of the conscious living being or even the advancement of human history!

Any personal belief or any personal point of view, any personal thought or any personal habit which does not bring more or true or complete knowledge of actual or physical reality or of actual or physical existence or of physical creation is nothing but simply a very heavy burden.

And that which is a heavy burden, a simple burden or not, has no real or has no true purpose.

In other words, that which is a burden is simply nothing but useless and a chaos maker or a reality destroyer.

Human reasoning or human understanding or simple point of view or scientific knowledge of existence or actual reality or physical creation thus far proves exactly the physical opposite of what is simply and actually expected from that very knowledge of existence or actual reality or physical creation or from human simple understanding!

Scientific knowledge of existence or actual reality or physical creation or human science really or actually has the very simple tendency or has the simple bad habit, for any habit is bad and useless, to really prove with what already is or with what already was with that that is but not.

But it simply and actually gets much worst!

Countless scientists or countless researchers entice multinational governments into spending enormous amounts of monies, enormous amounts of energies and even enormous amounts of irreplaceable natural resources to simply prove time after time the very exact physical opposite of what the countless scientists or the countless researchers actually or really wanted or physically expected.

Ironically, countless scientists or even countless researchers actually get paid very well for misinformation, for chaos, or simply for false or useless knowledge that simply bankrupt multinational governments or simply bankrupt global economy!

This is truly a lot of physical nothing for a lot of physical something!

That is the simple reason that the conscious living being simply cannot allow a wrongly used tool, or allow an idea or limited point of view that does not bring personal joy or does not bring a personal expansion of conscious mind, simply dictate conscious thought!

No star burns through the physical energy of another. No apple is physically tasted in another mouth!

Conscious Thought

What, therefore, is the simple purpose of conscious thought or the simple purpose of conscious reason or reasoning?

What, therefore, is the true or the real purpose of conscious effort or what is the real or the true purpose of conscious thought?

The single and the only true purpose of conscious thought or the single and the only true purpose of conscious reason or reasoning is to simply bring full or complete enjoyment or bring enlightenment of true or complete knowledge of existence or actual reality or physical creation, therefore bringing about more true knowledge of existence or actual reality or physical creation, and thus more freedom or more liberty of the human conscious mind!

In other words, the single and simple true purpose of conscious thought or of conscious reason is to simply bring to the human brain full or complete knowledge, full or complete knowledge here and now of human actual or physical surroundings and human personally interacting with the actual or with the physical human surroundings that would otherwise take countless of thousands of physical years to simply gather or to simply find or to even simply understand or to use and thus simply enjoy.

Gladness and joy or real enjoyment is actually or is really an expansion of the human conscious mind, but because the gladness and the joy or the enjoyment, or even a simple conscious thought, is used as energy thus the expansion does not last and the human conscious mind returns back to normal or to the original surroundings or original state.

Often, therefore, a simple thought is lost as energy through excitement, gladness and joy or enjoyment even before that simple thought was completed or acted out.

If the human being or the human race is therefore feeling a bit comfortable, therefore a bit of more freedom of human conscious mind!

But if the human being or the human race is simply feeling a bit uncomfortable or simply feeling a bit threatened, therefore less freedom of conscious mind or less freedom of conscious thought or less freedom of conscious reason, and even less physical freedom or less physical movement of body!

The human race is practically going back to the Stone Age, but not before the human race physically turns to stone!

Too often, therefore, a conscious thought is simply eliminated, rejected or simply forgotten or even ignored before the conscious thought is completed or fully understood or fully and simply enjoyed, creating confusion, misunderstanding or chaos to the human brain. Thus, the dream or the nightmare!

A dream or a nightmare is really or actually is a very simple continuation of an incomplete conscious thought or a simple continuation of an incomplete conscious emotion or half-truths or even a new point of view.

Most dreams or most nightmares are really created by the human brain simply trying to understand and to simply enjoy a new conscious thought, a new feeling or a new emotion or a new point of view that really or actually happened but was not completed or was not understood and thus was not fully enjoyed.

And since the human brain really has no previous or true or complete knowledge of that new reality or new creation or new point of view or the brain really has no real information to that new conscious thought, has no complete or real feeling or has no complete or real emotion, the human brain simply searches or simply rewinds the memory of the brain. Thus, the dream or the nightmare!

But that simple dream or that simple search turns into a lasting nightmare or a lasting mental chaos when the human brain simply tries to attach a simple and complete feeling or a simple and complete emotion from memory, but cannot find one.

And so the lasting nightmare or the mental chaos goes on and on and on until the human brain is fully satisfied or simply but stone dead!

The dream or the nightmare or the mental chaos is, therefore, the simple tool, the pipe or the monkey wrench deliberately thrown by the human conscious mind into the gear box of existence that really is the human brain!

Other dreams or other nightmares or other mental chaos is simply caused by physical pain or physical discomfort. For example, the very simple need for the human body to relieve or to empty the bladder while at sleep can really cause multiple dreams or nightmares!

Even a fever while at sleep can cause multiple dreams or even nightmares.

However, most dreams or most nightmares are simply and actually caused because of misinformation, false knowledge or half-truths or simply because of the need of the human brain to simply know or to simply complete knowledge of physical existence or actual reality or physical creation.

And the reality of the matter is that when the conscious living being has a personal understanding or has really learned something of reality, reality is no longer the same!

By the way, depending on what side of the body that is slept on, thus the dream or thus the nightmare or thus the mental chaos!

Sleep on the left side of the body, one type of dream or one type of nightmare. Sleep on the right side of the body, another type of dream or another type of nightmare.

The dream is as sweet or as bitter as the physical side of the body that is simply slept on.

The very simple reason for having different types of dreams or having different types of nightmares depends on what side of the body that is slept on is that the right side of the human conscious mind or the right of the human brain physically controls the left side of the body and the left side of the human conscious mind or the left side of the human brain physically controls the right side of the body.

In other words, the right side of the human conscious mind or the right side of the human brain, which is irrational or which is emotional, simply looks for information or true or real knowledge dealing with the emotions.

If no information, no recollection or no knowledge or no good or no meaningful feeling or no real emotion is found in memory, then the simple nightmare.

The nightmare or the chaos is the simple tool to try to explain an incomplete or a misunderstood human emotion or a misunderstood human feeling. The dream or the nightmare is the simple physical motion or point of view to simply try to understand an incomplete thought or an incomplete emotion or even a new experience or the reaction to that new experience.

Actually, a dream or a nightmare or a mental chaos is simply caused by a very simple denial or by half-truths or by false knowledge of physical existence or by false information. Even a simple lie can cause a lasting dream or a lasting nightmare or even a lasting mental chaos.

The left side of the human conscious mind or the left side of the human brain, which is rational or logical, looks also for complete information dealing with physical reality or with complete truths. If the opposite is true or is confusing, then the dreams or the lasting nightmares.

The dream or the lasting nightmare here is supposed to explain the irrational that happened in the rational or the rational in the irrational.

For example, the rational or the logical side of the human conscious mind or the human brain is simply trying to understand and trying to enjoy a physical action or point of view that really or actually happened, such as a real explosion, but the individual or the person consciously denied that physical or that real action or point of view through conscious disbelieve or not really believing that such a physical reality actually or really happened even though he personally saw the actual physical explosion, therefore really causing confusion or really putting the human brain in total chaos!

Another very interesting thing about dreams or nightmares or even mental chaos is that they are too often become multiple dreams or multiple nightmares or even multiple mental chaos in a single dream or a single nightmare or even a single mental chaos. That is, many dreams or many nightmares in one single dream.

Actually, a three dimensional dream or nightmare!

To the dreamer it was one continuous dream. But in reality, they were multiple dreams, at times more irrational than rational.

It is more like a dimensional dream or many points of view, where time stands still but many things are really or are actually happening at once and at the very same space or location!

This dimensional dream is the brain really searching or really analyzing multiple dreams or multiple points of view at the very same time!

The brain is really searching for a true or a real reason as to what made the conscious living being act or react irrational in the rational or act or react rational in the irrational.

In other words, since the brain cannot understand the new or the incomplete conscious thought or the new point of view or even a partial reality or a partial emotion or partial feeling, the brain looks or rewinds the memory in order to complete that new conscious thought, that new point of view or that new but irrational emotion or partial feeling or even a partial reality.

And so the lasting dream or the lasting nightmare really is equal to the reality acknowledged or to the reality already recognized plus the point of view or the simple understanding of that recognized reality minus the reality unseen or unrecognized or even the actual reality denied. In short, "D = +0-." Or simply put, plus zero minus.

Once again, the simple and the single true or real purpose of the human conscious mind is to simply but know and to simply but enjoy true or complete knowledge or to simply enjoy actual or physical reality or physical creation.

The true or the real purpose of conscious thought or of conscious reason or a much better point of view is simply to complete knowledge or to simply complete information of actual existence or actual reality or physical creation, here and now, and thus give more freedom or more liberty to the human conscious mind!

Not until the human conscious mind, as well as the human brain, is really complete or fully satisfied with the actual or the physical reality or the physical creation that really is here and now, there will be no more mental freedom or no more liberty of human conscious thought or no more conscious mind and no more liberty or no more physical movement of human body!

That is to simply say, not until the human conscious mind or the human brain is really complete or filled or fully satisfied with actual or with physical reality, there will not be more human conscious mind, no more knowledge or no more understanding or no new point of view and thus no more enjoyment!

That is the simple reason that humanity or mankind, as well as the human individual no matter how great or how smart, gets stuck and human civilization, as well as the human individual, crumbles to useless pieces time after time.

Knowing, simply understanding, and really enjoying reality, physical and actual existence or even physical or true knowledge brings about an expansion of the human conscious mind and more freedom or more liberty of physical body.

And that is the simple reality or the simple and thus the naked and the complete simple truth!

There is no real or true point of knowing, understanding or there is no point in having complete or having full knowledge of existence or actual reality or physical creation if nothing is really going to be done with that complete or perfect knowledge of reality!

True or perfect knowledge or true or real information is to know that "one simply but are!"

That is to really say, that one can simply be but more or but everything.

True or real knowledge of reality or creation or true or real information is to know that one simply exists and that one can really do something more about it!

To stand still or to stand motionless, therefore, is to have simply but existed and thus simply and actually just waiting to simply die once again, only this time through a lot of physical pain!

Wouldn't it be more than simply and wonderfully grandiose to simply know that mankind, really the conscious living being, but simply is the simple and the only actual answer or the simple and the only true or real solution to the final or to the ultimate question, to the final or to the ultimate problem or to the final or to the ultimate solution that has yet to happen?

Existence or physical creation created naked man but not mankind or not the human being. Mankind or the human living being consciously created or consciously completed mankind or the human conscious being when naked man physically broke all the laws of Nature!

The simple and the single true or real purpose, therefore, of the human brain is simply to complete or to make perfect existence, to complete or to make perfect physical reality or to complete or to make perfect physical creation or to complete the gaps of true or real knowledge!

That is, to acknowledge or to renew existence through a higher or taller consciousness which the brain will grant!

The simple and the single true or real purpose of the human conscious mind is simply the very same as the very simple and the single true or real purpose of the human brain, but with full understanding or with complete personal enjoyment!

That is really to say, the simple true or real purpose of the human conscious mind, through conscious thought or through conscious effort, is simply to help the human brain here and now to simply finish or to simply complete or to simply make perfect physical existence or complete or make perfect physical reality or complete or make perfect physical creation in and through personal enjoyment or in and through personal enlightenment or through personal acknowledgement!

So, it is a very simple question of how conscious living beings through conscious thought or through conscious effort and enjoyment will finish or will complete physical creation or finish or complete physical Existence!

By the way, true, real or complete knowledge is that knowledge that can physically be but proven because true knowledge simply exists!

Knowledge and Acknowledgement

Can you, the interested and the now glad and joyful reader, gladly and joyfully truly imagine what would had happened if the simple and the truthful or the real question as to the physical origin of the universe or as to the physical origin of all Existence, or even the truthful or the real question as to physical creation, were simply asked and simply answered when conscious thought first entered the human conscious mind?

Really, the physical world or physical creation would not be the physical and the mental lack and the physical and the mental chaos that the world or that physical creation really is!

There would had never been insignificant or useless ideas or useless points of view that, unfortunately, still today bring forth false, half-truths or useless knowledge, not allowing mankind to further expand his conscious mind into simple enlightenment or simply into completeness or perfection, and therefore, not allowing the future or complete or perfect Existence or perfect creation here and now.

Unlike the divisible atom, where the electrons cannot naturally return to the center or to the nuclei of the divisible atom or where the proton naturally cannot occupy electron rings or the electron field or shield, and certainly the physical universes would really be physically different and therefore all of actual Existence, the conscious mind or the conscious living being can deter, can reverse, can change or really can improve the conscious mind's or the conscious living being's point of view, believe or understanding or the way of conscious thinking or the way of conscious functioning or even the way the conscious living being learns simply by changing or improving conscious thought or even by improving mood or by simply acknowledging or by simply seeking to be enlightened or to be reborn!

By getting really interested or by improving one's mood is a very simple way to improve one's conscious thought or one's point of view.

One can see more by simply getting interested or simply by being glad or joyful or even excited.

Getting interested or getting glad or joyful or even getting excited is a major part of learning, but getting interested or getting glad or joyful or even getting excited is not an automatic process.

The conscious living being also has to make an effort to get interested or to get glad and joyful or even excited.

In other words, the conscious living being really has to make an unnatural conscious effort to learn how to really learn.

Once the conscious living being simply or actually knows or simply or really understands how to learn or really understands the real reason for really learning, which is really recollecting or simply remembering, the conscious living being will know or will understand how.

That is to simply say, once the conscious living being really knows or really understands why or really understands what or really understands the simple reason for getting interested or the simple reason for getting glad or even excited, the unnatural process of getting interested or the unnatural process of getting glad or even getting exited then becomes automatic, and the conscious living being through conscious thought or conscious effort will really or will actually know how (to complete Existence or how to complete actual reality or physical creation).

Interestingly, joy, gladness or happiness, as really is also conscious thought, is pure energy. A joyful, a glad or a happy thought can really or can actually keep the conscious living being a good number of days without sleep, without hunger and without feeling tired or stressed out.

Only when the joyful, the happy or the interested thought is gone or completed or has actually become natural, the conscious living being once again begins to need sleep, to need food and to need rest, both physically and mentally.

True, complete or real joy; true, complete or real happiness; or true, complete or real interest really brings to light physical things or the reality unseen.

True, complete or real joy or real happiness is not an automatic or is not a natural process at first, even though true, complete or real joy already exists.

And because mankind instinctually or naturally knows that true joy or that true happiness is or already exists, mankind seeks true joy or true or real happiness.

However, that is temporary or that is natural joy or that is temporary or natural happiness. That is the reason mankind really feels miserable or really gets into trouble looking for temporary or natural joy or temporary or natural happiness.

True, complete or real joy or happiness is unnatural, and therefore, really requires conscious effort or conscious thought.

However, this really has nothing to do with positive thinking. Positive thinking is only a very simple point of view. Positive thinking is not really or actually a good thing.

Positive thinking can really be a real hell or a very heavy burden. Mankind usually, as a creature of unnatural habit, imitates mostly what mankind really likes or what really brings mankind the most pleasure or the most joy.

Changing or improving conscious thought, changing or improving behavior or changing one's way of thinking or changing point of view is unnatural and is also not very easy. It really requires a lot of effort, a lot of energy and a lot of time for a lot of nothing.

Positive thought really is, therefore, one of those good sounding lies or good sounding half-truths or false science. Positive thinking bears all as it pretends to dress and nurture.

But once the conscious living being or mankind through conscious effort really acknowledges, realizes or really understands or really recollects the real or the actual reason for true, complete or real joy or real and lasting happiness, then true, complete or real joy or real and lasting happiness simply becomes an unnatural but automatic process as conscious thought once became an unnatural but automatic process as the very simple magnetic field or the very simple neutral shield always was an automatic process of physical existence or of physical reality.

In short, once the unnatural but automatic process of true, complete or real joy or real and lasting happiness kicks in, new energy is felt both in body and in mind. At first that new energy will feel like an ecstasy or a natural super high.

In fact, it is really or actually the first enlightenment! It is a true, a complete or a real joy or a real and lasting happiness that keeps coming, even without further effort from the conscious living being, leading to a second and final enlightenment or final acknowledgement. The very simple anticipation of the next enlightenment really or actually is what makes possible the new found joy or real and lasting happiness.

Ironically, when true, complete or real joy or real and lasting happiness begins automatically as once did conscious thought, the conscious living being really does not understand where it really comes from and so the conscious living being begins to fight real and lasting happiness, as the conscious living being once fought against conscious thought, by feeling odd or by feeling strange, and thus begins to reject that new and automatic and real joy or real and lasting happiness.

And in conclusion or in continuation, the most important thing in all of creation or in all of Existence, seeing or not seeing, once again, simply is or actually is but conscious thought!

Conscious thought only exist in conscious living beings. Conscious thought really is not a natural or is not an automatic process as is all of Existence or as is reality.

Even all of Existence or all of creation as well as non-existence really require something. Conscious thought is not a natural process and conscious thought really requires conscious effort from the conscious living being.

Even though knowledge simply exists or simply "the fact simply and already but is," the conscious living being does not know or does not learn or does not really recollect or does not really remember and thus simply cannot complete Existence or physical creation until the conscious living being makes a conscious effort to simply think and to simply ask that unnatural but true or real question which will simply bring the conscious living being complete or perfect knowledge or acknowledgement or enlightenment!

And the real or the actual purpose of complete or perfect knowledge is simply to let the conscious living being know or really acknowledge that the conscious living being but is and that the conscious living being can really do more with what the conscious living being could really be or could really recognize or could really recollect or could really remember!

Existence as knowledge simply is or simply exists; but ironically, Existence or knowledge knows not. However, just because all of Existence or all of knowledge really knows but not, really does not make Existence or knowledge dumb or useless.

All of Existence or all of knowledge acts or reacts through a simple magnetic or through an electronic impulse or acts or reacts through an actual magnetic instinct.

But knowledge does not know that knowledge really is or that knowledge simply exists. For example, the simple apple tree grows knowing not; but the tree's natural instincts, through knowledge already gotten or already present, is to produce a seed or to simply multiply itself.

The apple of that apple tree ripens or temporarily becomes almost complete not knowing that it does, so that the seed in the almost complete apple can be carried away, but not by the wind or not really by rolling on the ground even though the apple is almost round, but really carried away by living beings. That is the very simple or the actual reason for the sweetness or the sugar in the almost round apple!

The simple apple in the belly of the living beast or in the hand of the living being is simply or is really a continuation, and not the conclusion, of something that already was.

And even though the apple tree which is really but simple knowledge was really before the seed, the purpose of the apple tree was always the seed, an extension or an expansion of physical knowledge that really already was or was already present!

Another very interesting thing about the knowledgeable apple seed is that the apple seed not only carries all of the physical information of the apple tree, but also the apple seed carries all the knowledge or all the information about the environment or the physical surroundings where the apple tree was or is.

The knowledgeable apple seed no longer has to experiment with the environment or with natural surroundings since the knowledgeable seed already has that physical information. Instead, the seed as a new or as a second tree will gather more or new knowledge for the next apple or the third tree not yet present.

Whether the new or the second apple from the new or the second tree is sweeter or is more sour than the very first or the previous apple, simply depends, of course, on the physical reaction or the interaction that the living beast or the living being gave to the very first or to the previous apple.

The knowledgeable apple is often colored red to attract the eye of the living beast or the living being with hands, and the redder or the more reflective or shiner the apple, the better for the apple tree and thus the better for the knowledgeable apple seed.

If the natural or the physical surroundings are all red, then the knowledgeable apple with time simply changes to another but more attractive or more attracting color.

By the way, animal seeds or animal cells as do plants seeds have the very same ability or instincts to carry knowledge of previous generations.

But because of the animal's irrational or emotional behavior, or even point of view or new physical location, that first or that previous knowledge or instincts is rarely re-used and usually is forgotten, rejected or even lost.

A very good or an excellent example of lost knowledge or lost instinct is the domestic animal. Left alone in an unfamiliar environment the domestic animal will not survive.

Even house plants lose the ability to survive because of the lack of information or the lack of knowledge of the new environment or new surroundings.

Once again, conscious living beings are really the most important thing in all of Existence, and that is what Existence really reflects or that is what all of Existence really wants or really requires the conscious living being to know.

But not only that!

All of Existence also requires, also wants, or also needs, as the apple tree, the conscious living being to simply act or to react to knowledge, so that all of Existence as well as the living conscious being can be further completed or perfected through conscious thought.

In a way, conscious living beings are actually or are really inside a very large mechanical or infinite magnetic brain or inside a very large electronic brain; and conscious living beings are really the thinking brain cells or living knowledgeable seeds or even living knowledge processors.

Ironically, Existence will not be complete or will not be perfect until all the needs of the conscious living being are simply but met through conscious thought. And that is the actual reality!

The only true or the only complete or the only perfect or real beginning in Time and in all of Existence or even creation is the beginning of conscious thought.

The only limits that the human race, the conscious living being or that mankind really has are the very same limits that the human race, the conscious living being or that mankind really gives or really sets upon himself.

So, therefore, let's get glad and joyfully continue to think and rethink!

Thinking or conscious thought will one day keep the universes or creation as well as mankind from collapsing.

Conscious thought will be the conscious shield that will keep the simple universes or creation from collapsing or from imploding into a giant black hole or into a giant grave.

Conscious thought will be the conscious shield or the conscious energy that will one day really keep the simple human conscious mind or the human brain from physically collapsing or from physically imploding also into a hole or into a grave, thus saving the physical universes or saving creation and thus completing or really perfecting physical Existence or physical creation!

The final reality of the matter is that as long as there is conscious thought in the physical universe or physical universes, the physical universes or physical creation will not collapse or will not implode or will not even turn on itself.

The reality of the matter is that there was only one physical creation or only one physical beginning, actually a physical continuation of what already was.

As long as there is conscious thought, only one physical beginning or only one physical continuation or only one physical creation will be so until infinity!

Overcoming through one's Conscious Mind

One can only overcome through the conscious mind and never ever one can overcome through being dead or never ever one can overcome through death or never ever one can overcome through what another had said that he did for one so that one did not have do and neither one had to die because supposedly the other already has overcome death and one only has to die once or die that first time to triumph for other that is no longer nor is able to speak for himself.

But the truth is that just as one did alive before birth to alive be born to the world even though with a new form, in the world one also does before rebirth or before dying to alive one in the world continue but continue also as if in a new form because of the state of mind in where one truly will be.

But if one dies, thus truly will be the end of everything for one!

The sword as the shovel makes the space or makes the emptiness, but peace truly is made by the conscious mind.

And when one has peace, one has knowledge or understanding; and when one has knowledge or understanding, one has also the gladness and the joy; and when one has the gladness and the joy, one has also the abundance of life in life.

Now, one feels alive with every action of life but no one feels dead or feels death with every action of death. Life is about every feeling, be it good or be it bad.

And death is about nothing for coming to nothing or because of not receiving the knowledge or the acknowledgement to be reborn and continue with life but with life in complete gladness and in complete joy and also in complete abundance of real or true things that always were but that could not be seen until the rebirth of one through knowledge or through acknowledgement also of one.

The greater part of humanity has not come to know all the things that are in the word or in the very life of man and many have died without ever feeling or without ever being able to see the very minimum of the reality that is but has not been seen.

But a certain part of humanity had been illuminated or had felt the reality that is but that has not been seen, but that fortunate part of humanity or those very fortunate men could not enter into the presence of the reality that is but has not been seen because humanity or those fortunate men did not know or did not understand of what was based that reality or as to the reason for her.

On other words, many truly had felt or had known that there is more in reality but they had not known as how to enter that part of reality which is much greater than the reality which is or that is seen by the human eye.

And because of the lack of that knowledge of how to enter into a reality which is but that is not seen, those fortunate man created their very knowledge or ideas to be able to enter into the reality not seen and they began to name or to rename the reality no seen as the hereafter or after life and that one could enter her after death. What minds so curious but what minds so vain!

And all of those fortunate men died without ever knowing or seeing the reality and even less entering her, what a shame!

But when one makes the reality manor, thus one truly fails or to one the truth is missing with all sadness and with all anguish and also everything is as if zero or vanity for one even though one is one!

Now, just as the eye truly has its grandiose purpose to see and to know, thus the conscious mind also has its grandiose purpose of recognizing to be reborn!

And he that is reborn, thus he truly enters alive or enters in life to the reality not seen!

Now then, the reality not seen was always one but one truly reborn or truly renewed or one truly born again!

And when one is reborn or one is renewed or when one is born again, the reality not seen truly becomes much greater even though for being greater was not seen by one and reality will be truly seen by one in life because the reality not seen was always one reborn or renewed or born again!

To complete, truly fortunate is he that as of yet continues alive because as of yet he has the very grandiose opportunity of truly be reborn or of truly be renewed or of truly be born again and of finding himself in the reality not seen but it is felt and that it is forever!

The Paradise is one

The paradise truly is one and the paradise truly is for one because of what one has done for the paradise and one returns or one once again enters into paradise which is one when one truly understands or learns to multiply, to divide, to add and to subtract one at the very same time for much more so that one for much more can continue with complete gladness and with complete joy and one also can continue in complete abundance in existence without ever stopping from living or without ever stopping from existing in her...

Existence as everything which in her exists is composed of knowledge be it positive or be it negative or be it neutral.

And that knowledge also is composed of positive numbers, negative numbers and even neutral numbers as the zero.

Matter or light for also being knowledge is composed of positive, negative and neutral numbers, more like + 0 -...

In other words, E = + 0 -; where E is existence and + is light or matter; 0 is the neutral point and - or negative is lack or darkness or space.

And it can also be said like so, 1 = + 0 -; because existence is one or existence is a magnetic and the neutral part or the zero part or point can be transformed into positive or into negative or even into both.

In the case of existence, for example, here the zero or the neutral part is the universe. And when matter or light enters into the universe thus the universe has converted into a positive part, but when there no longer is matter or matter has stopped from being positive because of losing her energy in the universe thus the universe has converted into a negative part but the universe will not be very long in the negative because the universe will return to neutral or to zero...

Now, one is equal also to positive, to neutral and to negative. That is, one also is a simple magnet and as the same as a simple magnet one can attract, repel or even increase from one side or be it from positive to negative or from negative to positive.

One also could increase the neutral side by just leaving the negative side, but that will not increase the positive side. The positive side of one can only be increased through positive knowledge or through a positive identity...

In other words, one for being knowledge one is positive but one is in an environment where there is lack of knowledge and that lack of knowledge thus obviously is negative.

In the case of existence, existence is composed of four sides. One side is positive such as light or matter; and three negative sides such as darkness or the vacuum of space and space-time which makes the other two sides.

The attraction between the negatives sides and the positives sides creates a point or a neutral space or point zero.

That point is called the universe for having all the qualities of the four sides even though temporarily, temporarily because the universe changes from positive, because of the light entering into the universe, and the universe changes to neutral because of the light losing her light or her energy and what remains is dust; and then the universe comes to be negative because of there no longer being dust in the universe because the dust was removed by the black holes and because of there no longer being dust or matter in the universe thus also the black holes disperse, that way returning the universe to zero or to a neutral point and ready to receive new light or another beginning...

Thus, obviously, the purpose of one is to come into a greater positive state or of identity or of knowledge or of acknowledgement, more like to the right side!

And with every positive state or of knowledge or of greater identity which one enters thus there is an expansion not only of conscious mind but also there is an expansion of life or of time.

Now, here positive has nothing to do with positive attitude or action but with doing to have positive or useful knowledge or to have a greater identity which will bring to one a greater state of life and of conscious mind and thus eliminating all negative knowledge or thought and be able to continue with life as if a new life and in all abundance or in all plenitude as if on the right side of existence because now the neutral side of one has converted into a side of abundance or of plenitude, more like a side of energy which never ends...

When light enters into the vacuum of space or enters into the universe, the light enters + 0 -, all the opposite sides are united.

That is to say, all the opposite parts of light are united into one by the neutral point. The positive is united to the neutral and the neutral also is united to the negative as a magnet.

Here the neutral point is greater or double the size of the positive and of the negative but when light enters into the vacuum of space or into the universe thus the vacuum of space compresses the light and the light explodes causing the separation of the three parts of light.

The positive remained with the neutral while the negative became separated and began to float or spin around the positive and the neutral.

But the neutral now is less or manor because it was lost as energy and space also surrendered the very same amount of space which the light took...

In other words, when light enters into the vacuum of space or into the universe, light enters in the form of an element and that element is composed of three parts as a magnet.

And the vacuum of space or the vacuum or emptiness of the universe compresses that element or that magnet and that element or that magnet explodes into pieces in where the negative part takes position away from the positive part but the negative part stays floating or spinning around the positive and neutral parts because of the magnetic field or the electron rings which now were formed around of what used to be a magnet or a solid element...

Now, when element number one enters the vacuum of space or into the universe, that element or that light or that matter also has the weight of two.

That weight of two is so that that element can come to or can react as if element number two or can come to the next state of two. And if that element comes to that next state of two thus now its number would be of two and its weight would be of four.

That is, that light or that matter or that element one because of its weight would increase its positive side and would be able to react as if another light or as another matter or even as if another element and that way also expanding the time of that light or of that matter or of that element.

But that does not stay like that, because that element can continue for a long time making that same process up to 118 times or until coming to the element or state 118 which its weight would be 236 times, returning once again to one with its weight or with its potential of two!

And if the light or the matter or the element without having consciousness or without having idea that it exists has that grandiose ability of changing its state into a greater state to continue as new in the vacuum of space or in the universe, how much more we conscious beings also have that grandiose ability but we can enjoy that grandiose ability in all gladness and in all joy and in all plenitude!

Now then, before continuing I would like to make clear that not only one element enters the vacuum of space or in the vacuum of the universe but also at the very same time a grand total of 118 elements enter.

That is, that there enter 118 elements with their double weight. Element one has double weight and element number two has the weight of four and element three has the weight of six and that pattern follows until element 118 which its weight is 236...

Thus, one is the paradise as the paradise is one. But if one wants to come to be the paradise thus for the paradise one must do. But if one does not want to be the paradise thus simply do nothing and death will make for one because the paradise or immortality is not obligated.

But he that has died or that has let death herself do for him for he doing for dead or wait for death as if death were some glory, thus he has lost the very grandiose ability of transforming into his very own paradise!

Now then, according to the knowledge which one brought into the world thus with that very same knowledge one will be confirmed into the world.

If one brought into the world knowledge of dead, as dead one will be known to the world and as dead one will be thrown from the world and there will no longer be memory of one...

But if the knowledge which one brought to the world was of life alive thus with that very same knowledge of life alive one will be confirmed to the world.

In other words, because of one being born alive into the world and the world confirm one born alive thus one was born as contender because one will continue contending not only for the peace of life but also one will continue to contend for the knowledge of life until that peace and that knowledge is achieved in the form of consciousness or of knowing the identity of one as a conscious being...

Once one has come to consciousness thus also one has come to be justified or be reconfirmed as contender even though no one else has said so.

That is, not until one seeks for that justification or for that reconfirmation from above from the very heavens thus one will not be justified or will not be reconfirmed.

And that justification or reconfirmation is done because of one presenting oneself for more to the heavens, more as to God or more as to the Creator or even more than Creator.

In other words, one is confirming or knowing that there is more without one as of yet being able to see it or knowing it because what is before one is for more or for double of one...

But one does not have to be religious or have any faith because one truly is reconfirming or one is giving or presenting knowledge as if to the double or the twin of one, only that the power and the authority of taken one to a greater state is with the double or the twin of one!

And if one's presentation or confirmation or knowledge was accepted thus one will be told instantly be it with a voice or with a lightening thought for being instantaneous and one will be filled with gladness and with joy and one will be a very glad and a very joyful contender that truly has entered into a greater state or identity as contender beloved or servant beloved of God or even as savior beloved of God, depending on one's presentation!

But if there was no voice or if there was no lightening thought thus the presentation of one was not accepted because it is not a presentation which truly reflects the knowledge of one or which reflects what one knows or what one has reconfirmed, because one will be responded to with the very same knowledge which one presents...

Once one is responded to thus one enters into a greater state or identity even thought that new state or that new identity as servant beloved or as savior beloved is a neutral state or identity, more like the state of desolation because of not knowing what to do next and because of losing the gladness and the joy which were before entering into the new state.

And if a voice calls to one and by the name of one in a tone of two repetitions as if a distant voice and as if unknown, even though she already was heard by one more than once, thus it would be to grant to one some request of one coming out or of exiting the desolation if it is that one responds to the voice...

But if one does not respond to the voice thus the voice will be quiet and could be very quiet for a very long time and that way expanding the time of the desolation.

But even though the voice may become quiet for a long time, one cannot remain quiet because the salvation of one or the entrance of one into a greater state depends on that voice because that voice is what reconfirms the exit or the entrance into a greater state or identity such as the state or the identity of savior beloved of God or of dwelling beloved of God or even of plenitude beloved of God or the last state or identity of the right side of God...

Thus, one has to open the mouth of one because just as one through the mouth of one saves the body of one thus one also through the mouth of one saves the life of one with the simple request of salvation.

And even though one may receive salvation with all the power and with all the authority of the holy heavens, one must go on requesting until coming to the right side of God or to the positive side of existence, which truly is one converted to the right side or to the positive side of existence, more as to the paradise of one...

The human brain with the conscious mind

The human brain alone with the conscious mind is really the only organ in all of existence that can control space and transcend through time and other dimensions.

The human brain alone with the conscious mind will one day keep the universe as well as the other universes from collapsing.

The human brain alone with the conscious mind is also the only organ in all of existence not only capable of knowing or understanding it exists, but also it is capable of healing itself as well as everything else.

Now then, when a person receives a sudden impact in his or her very valuable life and that sudden impact separates the outside world from the inside world and that person survives the sudden impact, that person is no longer the same person.

The worlds, both in and out, are not the same. Most people come through and when then do, they are much better and they see both worlds with a better or a new perspective.

Changing a condition or situation is very difficult or next to impossible. Staying with the flow is easier and practically less resistance. The path of less resistance is more attractive, even if it is the path to nowhere or to doom!

Also, change is very personal, very emotional and very demanding. But change or improvement is only a continuation of things to be.

Change is an improvement of the same. Change really is a simple improvement of the same thing. Change is not changing one thing for another, but really improving a condition or situation.

How do we improve our condition or situation?

First, we must admit or know or understand that we have a problem because the first solution to any problem is knowing or understanding that we have a problem!

Do we have a problem?

Secondly, the only requirement in change is desire. Desire is power! If we really desire to improve, then we will definitely improve!

We have to have that simple desire to improve or change because that is what starts the healing or improving process. Not too many medical doctors, not even good ones, will tell us this very simple but very significant and thus very powerful secret.

Believing is thus creating. Belief begins to create and that creation makes change not only in our brain and mind, but also in our physical body and the outside world.

Without a Higher Consciousness

If one came out or if one were thrown out from the womb of a man into the womb of the world, one would die even though in a multitude.

If one came out or if one were thrown from the womb of a woman before one was formed, one would die.

And if one came out or if one were thrown from the womb of the world into outer space, even though a man and a multitude of forms, one will die!

Because in truth, without birth nothing is born! And without rebirth, nothing survives or nothing lives on.

But once one is reborn or one is renewed, the earth is as if a new earth and to leave it is as useless as emptiness because now the earth is more abundant because of one's rebirth.

Now then, rebirth has nothing to do with dying or nothing to do with death because rebirth truly has to do everything with life as birth has everything to do with life.

And death is because of the lack of life, because of the lack of the continuation of life or because of the lack of the expansion of life.

But that lack of life is caused by one when one no longer contends for the knowledge of life as one once contended for the knowledge of life before one's very birth.

But once one overcame before life, life one received and took the form and once the form of life was complete, thus into the world one was born.

In other words, one contended inside a woman to receive the other part of knowledge that would allow one to take full form and thus be born into the world.

And once in the world, contend once again in the world for more knowledge to survive in the world.

But to be reborn even though in the very same world or in the very same times, one must contend not only with man but one must also contend with the very heavens as one contended in the very womb with others to overcome and one overcame and one received knowledge of life to be born in life!

But unlike the womb of a woman in where only one overcame, in the world every man has the very grandiose opportunity to overcome for the knowledge of rebirth and knowledge of rebirth or knowledge of salvation to receive knowledge of rebirth or knowledge of salvation to be able to take the form of rebirth and thus be reborn!

Now then, the true knowledge of rebirth or the true knowledge of salvation truly comes from above from the very heavens and just as one presented oneself with true knowledge in the womb to true knowledge receive or to be complete with that true knowledge to be able to take the true form of life, one must also present oneself with true knowledge or acknowledge the heavens so that the very heavens can complete one with the true knowledge of the heavens or with the true knowledge of rebirth or with the true knowledge of salvation!

If the heavens do respond to one, they will also respond to one with gladness and with joy and one will also feel abundance or one feel a presence and that abundance or that presence will say with a thought or with a voice something but according to what knowledge or acknowledgement one presented!

If one was granted rebirth or salvation, one will be acknowledged so from above!

But if one was made a grandiose promise instead by voice from above then one must wait until one is ready or one is able to receive the promise or the reward.

But the promise or the reward was not specific but it has to do with one and with the very heavens and in life.

Also, once the heavens have truly spoken to one using a thought or a voice, one has become beloved or servant beloved of the heavens.

And if the heavens call to one but one does not respond, the calling or the voice or even the thought will stop and visions will begin.

But if the visions frighten one for being so real or so true, the visions will also stop and dreams will begin.

The dreams, however, will be according to the knowledge one already knows but too often one knows not what one knows and the dreams, which are really messages but only for one, can be very confusing or even meaningless to one!

Some dreams can be dreams of consul or comfort while some others can be to justify one as contender or even as servant beloved!

But when one is ready to receive the acknowledgement of rebirth or of salvation or the promise or the reward, one will be called once again and it would be very beneficial for one to respond!

Be, however, very careful if when one is asked to request for something because one will get it and it may not be as one actually thought or believed because it will truly be according to the will of the heavens!

Therefore, request for the salvation of God or for rebirth or a higher mental consciousness because in doing so thus one has revived or one has truly given rebirth to God and one has truly become or one has overcame as savior beloved of God, the greatest of rewards or the greatest of promises!

The Need for True Faith

True faith is to believe that there is a higher mental consciousness and that one can really achieve that higher mental consciousness through the action of oneself as one did before birth.

A conscious being or a man with faith truly has a very grandiose advantage over a conscious being or a man without faith or without true faith because that conscious being or that man truly has a knowledge which does not have that conscious being or that man without faith or that do not have true faith.

True faith is knowledge but not all knowledge is true. Faith also is belief or is believe but not all belief or not all believe is true because to come to belief or to come to believe it is truly through the act of oneself such as being born was an act of oneself and that act of oneself was to do to be born.

Thus, one truly is not born by instincts or by another. One truly is born by doing by oneself after one comes out with all gladness and with all joy from the grace or from the womb of a living being and one enters into the womb of another living being.

And when one enters with all gladness and with all joy into that new grace, one makes the movement to find knowledge of life and in that knowledge of life one enters with all gladness and with all joy and one also is formed or is transformed into that very same knowledge of life.

Once that one has transformed, one takes the position of humiliation or one humbles for much more to come out or to be born to the world. And once in the world, the world knows one with the very same knowledge that one brought into the world.

That is to say, if one was born male, male one will be known by the world. And if a female was born, thus she will be known as a female.

But in the world, in where one once again must do to have or to find true knowledge, there are many believes that the world forces one to take or to accept and not for being true or facts but rather for being customs or for being part of the society where now one was born.

And because of those believes or that knowledge thus one stops from doing or from seeking to have true knowledge or true faith to once more take the form of that knowledge or the form of the faith or of the belief.

And since one does not take the new form of true knowledge or of the true faith, thus the form or the body of one begins to decompose or to lose the form until one dies and because of the death of one there will no longer be more form of one.

In other words, death truly is because of lack of true knowledge or of true faith in the world!

Thus, just as darkness is for lack of light and just as the cold is for lack of heat and just as space or emptiness is for lack of something or of matter, death also is for lack of true knowledge or of true faith in the world because the world truly is as the womb of a woman that one also has to overcome as one truly overcame in the womb of that woman to receive knowledge of life to take that form of that knowledge of life and one be born or enter into the womb of the world in that very form or with that very same knowledge.

And once one was in the womb of the world, one would complete the form of one as contender or as a man or as a woman with true knowledge or with true faith and one once again take the form or take the position of humiliation or the form or the position of humbleness for more, but this time for the form of new life or of revived life or the form of salvation as savior beloved to continue with life!

But that new life or life revived or life as savior beloved truly is a life in complete gladness, in complete joy and also in complete abundance because nothing will ever lack here on the earth and the earth will be as if in the very heavens and the heavens will be as if others or new heavens because of the earth truly becoming as if another earth because of the earth truly becoming as if in the very heavens.

Thus, a man or a woman with true knowledge or with true faith has the very grandiose advantage over another man or over another woman that does not have true knowledge or true faith because that man with true knowledge or with true faith has also the very grandiose opportunity of reviving or of being reborn and of continuing with life while the other man without true knowledge or without true faith will die and will stop from being for all the times.

Interestingly, that one does not have to complete or that one does not have to wait that the form of one becomes complete as man because once one is able to think or one is able to understand or one is able to speak for oneself, thus one can do for the true knowledge or for the true faith which will truly give one the new form or the reform of savior beloved with all the power and with all the authority of the heavens here on the very earth…

Also, with the true knowledge or with the true faith comes to one a new identity and with that new identity come or arrive better or new thoughts or thoughts filled with wisdom of heart or a new way of thinking.

That is to say, that one truly enters into a better state of thinking or of identity and also of one feeling glad and joyfully in abundance or abundant.

And with every state or mental consciousness that one comes to enter, thus there will truly be another state or there will truly be another form of thinking and of feeling or of identity.

The plus Zero Negative Factor

(+) (0) (-)

Those that accept an idea blindly or without trying or testing that idea, thus they unknowingly become liars.

But nonetheless, liars they are and as liars they live and will lie to others to convince them to accept or believe and they will keep on lying until death herself and death herself will close their lying trap.

Now, when one does the movement or when one gets interested for knowledge, thus one truly becomes that knowledge.

When the universe came to exist, the universe came to exist because of knowledge. The proof is in the numbers or in matter which is composed of elements and the elements in turn are really numbers, real numbers.

In other words, matter is knowledge because matter is composed of elements and the elements are composed of numbers and the numbers are composed of positive, neutral and negative states, thus the, (+ 0 -), plus zero negative factor.

What this means is that knowledge can become neutral or void or useless and then become negative or contra productive if at first nothing was done with that knowledge, such as converting knowledge into acknowledgement or a positive or useful respond.

In the same manner as above, the universe or the vacuum of space becomes positive when matter enters the universe or the vacuum of space.

But as matter begins to burn out, the universe or the vacuum of space begins to turn neutral until it turns negative, negative because of the black holes which now rule the universe or the vacuum of space which they now also suck up any remaining dust or matter to make space for another beginning.

But this new beginning is as if the very first beginning because there will not be any trace that there was ever a first beginning.

But the above matter would only be a theory or an idea if the conscious being, which is also knowledge, did for acknowledgement and the conscious being would have the power and the authority over matter to refresh matter and thus keep the universe always positive and refreshing.

Now then, acknowledgement or reconfirmation from above is really one's brain offering to one a higher or taller mental consciousness and if one accepts thus one will really have a higher or taller mental consciousness and one will see things as if they were always new.

Summing up the brain, All knowledge is beginning as also all acknowledgement is to renew or is to make as if new or is to revive that very beginning, but that very beginning will become as if two because the very beginning or the original knowledge has not stopped from being, but rather, that everything has become as if new and because of that newness or renovation, thus everything known or what was known will be seen, because it is, five times the abundance or the quantity.

In other word, according to the good knowledge which one gives to the things, thus the things will be to one because in truth one is the beginning, not only of one, but also one is the beginning of many more things by one being born alive.

Thus, obviously, the greater the knowledge that one has of oneself, thus the greater the knowledge that one can give to the rest of things. For example, when one is born alive, thus good knowledge of life one also brings into the world.

And because of the good knowledge of life, thus now a man is father, another man is grandfather and another even brother or uncle of the one born alive and who brought knowledge and also beginning and even continuation of life.

And also, now a woman is a mother, another woman is a grandmother and another woman is even a sister or even an aunt of the one born alive and who brought knowledge and beginning and even continuation of life, continuation because all the other persons in the family were also reborn with that birth alive of one.

But if one were born dead, then one would have not brought good knowledge of life into the world and there would have not been beginning or continuation of life!

Thus, no one was reborn through the dead of one!

Thus, as we have seen above, the good things are according to the good knowledge that one gives and because of that very same knowledge, not only the things react but also one reacts with the things.

That is to say in truth, to know or to name a woman or a girlfriend as princes, that makes one also a prince.

To say beloved to another or to a son because of the love which one feels in truth, also makes one beloved.

Now then, the very same thing is with God or with the human brain!

That is to say again in truth, when one gives knowledge to God or the brain as Creator, thus one has revived God or the brain as Creator and as Creator, God or the brain will present to one in one form or in other but one will know that it is God or the brain even though God or the brain does not say it is God.

Now, after the loving presence or the loving grace of God or the brain, thus one will be beloved of God or of the brain.

That also makes God or the brain beloved of one and that can make God or the brain to call one. God or the brain does not say that it is God or the brain calling because God or the brain believes that one has already that knowledge because of the knowledge of God or brain Creator which one has given to God or the brain!

So then, if one responded not and if one did not hear the voice again which called one by name, by one's name but one did not respond because one did not recognize the voice, then both are in desolation, in desolation because both are beloveds of each other!

And where there appears beloved or beloved of God, thus also there appears desolation because God afflicts his beloved or the brain afflicts the conscious mind!

And God the Creator or the brain afflicts his beloved through desolation and with lack of consul or with lack of knowledge or lack of acknowledgement.

But God or the brain also is in desolation through the desolation of one or through the desolation of beloved of God or the conscious mind.

Thus, what has to be done in truth?

Thus in truth, while one as beloved of God continues in desolation, then God or the brain also is!

And what is in truth desolation?

In truth, desolation is lack of good knowledge or lack of beginning or lack of continuation. In the very beginning one gave to God or the brain the good knowledge of Creator and that made one part of God, one as creation or one as servant beloved.

But since one did not give more knowledge or continuation, then both parts returned as if before the beginning, where there was no knowledge, but this time is the affliction, which is to know that there is lack because once there was and it was felt or known in abundance!

And so, one now has to revive the beginning with more knowledge or acknowledgement!

Just as once God or the brain was revived as Creator through knowledge of one toward God and that made one also creation or servant beloved of the Creator, thus now one has also to know or one has to give greater knowledge or acknowledgement to God Creator!

That is in truth, God or the brain is Master and Creator and God or the brain can be a loving Master as also a loving Creator, but there is something much greater than that!

And only one alive is the one that can truly make that possible, even though God is a loving Master and loving Creator all mighty!

Thus, in truth, the much greater that there can be is Father, Father Beloved!

And the much greater that in truth that creation or that a servant or a beloved of God can do for the Master God Creator is or the brain to give the knowledge or the acknowledgement of Father beloved so that the Master God Creator or the brain as Father beloved gives one as creation or as servant or as beloved of God the good knowledge or the acknowledgement of son beloved of the Master God Creator or the brain, which in reality is a higher mental consciousness or really is an expansion of mental consciousness...

Summary

Existence simply is. Existence always was. And Existence will always be.

In short, all of Existence, physical or not, always existed forever and ever. Nothing or lack or nothingness is really also something.

Nothing, lack or nothingness is also a major part of Existence.

Actually, Existence is really made-up of three parts which really are not or three parts which simply do not exist as physical Existence really exists.

Nothingness, lack or simply non-existence also exists but non-existence does not really make possible physical Existence.

Non-existence or lack does not really or does not actually happen by itself and non-existence or simply non-reality or lack does not make possible physical Existence or does not make possible actual or physical reality even though non-existence or lack or nothingness is really or actually a reality.

That is to simply say or the simple reality is, non-existence or nothing or nothingness or even lack does not really happen by itself.

Non-existence, nothing or nothingness or non-reality or even lack really requires or really needs physical Existence or really needs or requires physical or actual reality.

In short, non-existence or nothing or nothingness or non-reality or even lack, therefore, really or actually requires or really or actually needs physical something or something solid.

That is, non-reality requires physical reality. And nothing, non-existence or lack is something in itself and, therefore, exists.

Once again, nothingness or non-existence or even lack is a reality, although not physical or actual reality.

However, even though non-reality or nothing or non-existence or even lack simply exists, nothing, non-reality or non-existence or even lack really cannot exist by itself.

Nothing or non-existence or lack needs or really requires something or nothing needs or really requires all of Existence to exist.

Physical Existence or physical or actual reality is really the simple product of Existence and non-existence. Physical Existence or physical or actual reality, therefore, completes non-existence.

That is to simply say, physical something or physical reality really or actually completes or really or actually makes perfect nothingness or nothing or makes perfect non-existence or makes perfect even lack.

Nothing, lack or non-existence is not complete or not perfect. Only physical reality or the naked truth can really or can actually complete or can really make perfect non-reality or can cancel lack by filling or by completing lack.

Physical Existence or physical something or physical reality or actual reality really makes possible non-existence or makes possible non-reality or makes possible lack.

Contrary to common or to popular believe, physical something or physical reality or physical existence did not come out of that that was not.

Physical existence already was or was always present at all times.

And of what already was present all the times, more came out or more came to be. Nothing or lack really or actually attracts that which already is and not what is not.

In other words, the simple shadow or the simple shade is not because of lack of light. Physical Existence attracts nothing.

Physical existence simply or actually or really attracts nothing or actually or really attracts non-existence.

Physical existence does not attract physical existence because physical existence is already complete or already full or physical existence simply is physical.

Only opposites really or actually attract. Non-existence or nothing or even lack really does not attract what is not. Non-existence or nothing or even lack really attracts what is.

Physical something or physical reality does not attract itself because something is already physical reality, but physical something or physical reality actually or really attracts that which is not.

Non-existence or non-reality attracts physical existence or physical reality since non-existence or non-reality is not physically complete or is not physically perfect.

Even though nothing or non-existence or lack is a reality that exists, nothing or non-existence or lack is a reality that lacks physical reality.

And lacking physical reality, nothing or non-existence or lack is, therefore, not complete even though nothing or non-existence or lack is greater in size than physical reality or physical existence.

That is to simply say, no real or true question no matter how small or how simple is truly or really answered unless first asked and then a truthful or a real answer is really expected.

The very simple question or point of view is really or actually recognizing or really remembering or really recollecting that there is but more!

Something or reality really goes into nothing or into lack, and not the other way around. Nothing or lack can be filled if it already is filled. Occupied space is not space. Occupied space is really the physical or the actual universe.

And if nothing or if lack already is filled, then it is physical something or actual reality and not lack or nothing.

That is to simply say, if it is filled, then it is not empty and not lacking. And such is the simple rationality of all physical or actual Existence or actual reality.

And so the closer to the rational, no more irrational to the conscious living being or to conscious thought.

Physical existence or physical something or even physical reality speaks through physical creation, but physical existence or physical something or even physical reality speaks with a double meaning which too often adds to physical existence or too often adds to physical something or too often even adds to physical reality through too much contradiction. So even actual or physical reality is a point of view or a way of thinking!

And so the more rational to some thinking being the more irrational to some thinking being. Nothing seems like it is because nothing is unseen.

Nothing is a reality not really seen. Nothing is the lack of physical reality.

For example, to go up is also to go down; to hang upside down is also to hang right side up; to pull is also to push; to leave is also to arrive; to come is also to be already present; to simply multiply is also to simply divide; to add is also to subtract.

As one moves forward one step, one has also one less step to move back or forward; and also to point one way is also to point to many ways or directions; to close one's eyes simply is not to stop seeing.

But because no rational understanding or no rational knowledge is really found in the irrational by itself, which is really a major part of the rational, the other percent of that that already is here and at once cannot be really or actually seen. Remember the animal mind and the tree?

Although the simple tree is entirely present and complete with fruit and all, the animal mind cannot see the simple tree because of the simple point of view of the animal mind.

Knowledge simply but is or exists. Knowledge is also physical or a physical reality. Knowledge as physical existence always existed and physical knowledge will always exist, even without conscious living beings.

Learning is, therefore, really simple recollection or actually or really remembering and remembering is really completing.

The greater the point of view of the conscious living being is, the greater the learning or the greater the simple recollection or the greater the remembering.

The greater the learning or the greater the actual or simple recollection or remembering, the greater or the more complete Existence or actual reality will thus be.

The universe and Time are simple products of non-existence and Existence. That is to say, the universe and Time are products of two simple opposites: space and matter, negative and positive, non-existence and existence, vacuum and expansion, black and white, that which physically already is and that which physically is not.

Where those two opposites physically meet or physically intersect or even physically divide, there is the physical universe and physical motion or Time.

Time, which is physical motion or is physical expansion, is simply created by the simple magnetic fields or simply created by the physical magnetic attraction of two simple opposites: space and matter.

The space cube and the matter cube or simply all of Existence is a simple electric motor, really a magnetic motor or generator.

The electric or the magnetic motor is made possible by a very simple magnetic field or a magnetic shield.

In other words, the simple magnetic field makes possible the physical motion in the motor. In the same way, the magnetic field of the two magnetic cubes makes possible motion and Time or expansion.

Time, like the physical universe and physical Existence, always was and will always be. Time, which is simply physical motion or simply physical expansion, had no physical beginning and will have no physical end.

The physical motion or the physical expansion in Time is created by the simple magnetic field of the two magnetic cubes.

Time, therefore, is a simple point of view. Any physical beginning is simply a physical continuation of a physical existence that already was.

To go in, for example, is a continuation from something that already was. There was not really a first or an actual physical beginning, but a physical continuation because physical existence already was.

The only true or the only real or the only actual beginning is, therefore, that of conscious thought. Conscious thought, therefore, really completes physical reality or completes physical or actual existence.

That is, conscious thought really or actually makes or really or actually completes physical existence or actual reality. Conscious thought really or actually makes possible the reality not seen!

There were 118 physical beginnings that all began at once with 118 endings that all ended at once.

That is to simply say, if something began and ended at the very same moment, can it be called beginning?

Can it be called end?

The beginning and the end are but the very same thing. They are both really a continuation, and they are both but a very simple point of view.

For example, when leaving, one is also arriving at the very same time even though the actual arrival or the physical point of view is much later or the actual arrival or the physical point of view really will be completed or perfected much later.

In other words, the actual arrival is a very simple continuation of leaving. Even aging or growing old is a simple physical or actual continuation or expansion or even a point of view.

Once again, when turning left, one is also turning right, turning south and turning north at the very same time and at the very same location since physical motion is simply a continuation. When falling or when moving down, one is also going up to the ground!

North, South, West, East are really different names to the very same thing or the same place and that place or thing is also present at the very same time and at the very same place or space; and those names are also different names to a very simple and the very same point of view.

There is no such thing as a true or as a complete or a perfect north, a true or a complete or a perfect south, a true or a complete or a perfect west or a true or a complete or a perfect east, no matter how straight the magnetic needle points. All of those points of location or space, once again, are really present in the very same physical location at the very same time.

That is to simply say, everything exists at once, but not everything is really seeing at once simply because the conscious living being limits his conscious thinking, simply limits his way of thinking or simply limits his point of view

According to the naked knowledge or to the understanding or even the simple recollection or acknowledgement that the conscious living being really or actually has of Existence, the universe or physical creation, Existence, the universe or physical creation will thus be.

In other words, the greater the point of view or the greater the recollection, the greater Existence or the greater the physical creation will simply be.

So in reality, Existence equals plus zero minus. Or simply put, "R = +0-." Reality or physical creation really depends on the point of view of the conscious living being.

Once again, yesterday, today and tomorrow are also points of view. They are really the very same thing and they are also present at the very same time.

That is to simply say, tomorrow is already spent and tomorrow is only a point of view. Yesterday is a future to another yesterday.

Actually, yesterday was many futures. And yesterday is still here today. Today is a future to many a past.

Once again, the simple reality here is that today is multiple futures to many a past. Tomorrow, though a future of many a yesterday and a future of today which is another future, tomorrow is also a past and tomorrow also is already spent. Yesterday, today and tomorrow are all really present at the very same time here and now.

Thus, reality depends on where the conscious living being stands. That is, actual or physical reality depends on point of view. And what is point of view if not but conscious thought?

Time or beginning is, therefore, a simple point of view. Time or beginning or point of view is a simple continuation of that which already is, of that which already was and of that which will always be, and that is all of Existence.

All of Existence really is present all at once and in the very same space or location. The most interesting thing is that when the conscious living being gets to that point or that location or space, where everything or all of Existence meets at once, is that nothing will really be seeing and that nothing, not even anything, will really make any sense. In other words, all of Existence in itself or by itself really makes but no sense.

A physical part or a physical piece has to really or actually disengage or has to really or actually depart so that the physical greater or the physical larger part or physical infinity can be seeing through the physical smaller or through the physical lesser part; and also that the physical smaller part or the physical finite can be seeing through the physical greater or can be seeing through physical infinity.

To simply count or to simply add, therefore, the conscious living being has to simply take apart or has simply to divide.

And ironically, really multiply and subtract all at once or at the very same time.

The only thing that the conscious living being should really know or should really understand or should really realize or simply or actually recollect about time is that the conscious living being is limited to time or limited to physical expansion or simply limited to point of view only because the conscious living being limits himself through conscious thought.

And that the conscious living being must, through conscious thought or through conscious effort, change or improve his point of view or change or improve his way of conscious thinking and thus be able to complete or to make perfect physical existence or complete or make perfect physical creation or the conscious living being simply dies, therefore, allowing for a second physical beginning or a second physical creation of the universe and universes.

However, even though a second physical beginning or a second physical creation will be twice the size of the first physical beginning or twice the size of the first physical creation with even bigger Big Bangs, that is no reward.

A second physical beginning or a second physical creation is no second chance. A second physical beginning or a second physical creation is really or actually the failure to complete or the failure to perfect the first physical beginning or the first physical creation (through remembering) or through acknowledgement.

By the way, those conscious living beings living and present here in this universe or the other present universes will not be present or will not come on the next physical beginning or to the next physical creation.

Every physical beginning or every physical creation is really a one-time deal and only one opportunity to complete or to make perfect that physical beginning or that physical creation and not any other physical beginning or not any other physical creation.

And as long as there is conscious thought in the present universe and present universes, that one time sensitive deal and that one time sensitive opportunity is on.

Conclusion

Can you, the interested and the now glad and joyful reader, gladly and joyfully truly imagine what would had happened if the simple and the truthful or the real question as to the physical origin of the universe or as to the physical origin of all Existence, or even the truthful or the real question as to physical creation, were simply asked and simply answered when conscious thought first entered the human conscious mind?

Really, the physical world or physical creation would not be the physical and the mental lack and the physical and the mental chaos that the world or that physical creation really is!

There would had never been insignificant or useless ideas or useless points of view that, unfortunately, still today bring forth false, half-truths or useless knowledge, not allowing mankind to further expand his conscious mind into simple enlightenment or simply into completeness or perfection, and therefore, not allowing the future or complete or perfect Existence or perfect creation here and now.

Unlike the divisible atom, where the electrons cannot naturally return to the center or to the nuclei of the divisible atom or where the proton naturally cannot occupy electron rings or the electron field or shield, and certainly the physical universes would really be physically different and therefore all of actual Existence, the conscious mind or the conscious living being can deter, can reverse, can change or really can improve the conscious mind's or the conscious living being's point of view, believe or understanding or the way of conscious thinking or the way of conscious functioning or even the way the conscious living being learns simply by changing or improving conscious thought or even by improving mood or by simply acknowledging or by simply seeking to be enlightened or to be reborn!

By getting really interested or by improving one's mood is a very simple way to improve one's conscious thought or one's point of view. One can see more by simply getting interested or simply by being glad or joyful or even excited. Getting interested or getting glad or joyful or even getting excited is a major part of learning, but getting interested or getting glad or joyful or even getting excited is not an automatic process.

The conscious living being also has to make an effort to get interested or to get glad and joyful or even excited.

In other words, the conscious living being really has to make an unnatural conscious effort to learn how to really learn. Once the conscious living being simply or actually knows or simply or really understands how to learn or really understands the real reason for really learning, which is really recollecting or simply remembering, the conscious living being will know or will understand how.

That is to simply say, once the conscious living being really knows or really understands why or really understands what or really understands the simple reason for getting interested or the simple reason for getting glad or even excited, the unnatural process of getting interested or the unnatural process of getting glad or even getting exited then becomes automatic, and the conscious living being through conscious thought or conscious effort will really or will actually know how (to complete Existence or how to complete actual reality or physical creation).

Interestingly, joy, gladness or happiness, as really is also conscious thought, is pure energy. A joyful, a glad or a happy thought can really or can actually keep the conscious living being a good number of days without sleep, without hunger and without feeling tired or stressed out. Only when the joyful, the happy or the interested thought is gone or completed or has actually become natural, the conscious living being once again begins to need sleep, to need food and to need rest, both physically and mentally.

True, complete or real joy; true, complete or real happiness; or true, complete or real interest really brings to light physical things or the reality unseen.

True, complete or real joy or real happiness is not an automatic or is not a natural process at first, even though true, complete or real joy already exists.

And because mankind instinctually or naturally knows that true joy or that true happiness is or already exists, mankind seeks true joy or true or real happiness.

However, that is temporary or that is natural joy or that is temporary or natural happiness. That is the reason mankind really feels miserable or really gets into trouble looking for temporary or natural joy or temporary or natural happiness.

True, complete or real joy or happiness is unnatural, and therefore, really requires conscious effort or conscious thought.

However, this really has nothing to do with positive thinking. Positive thinking is only a very simple point of view. Positive thinking is not really or actually a good thing.

Positive thinking can really be a real hell or a very heavy burden. Mankind usually, as a creature of unnatural habit, imitates mostly what mankind really likes or what really brings mankind the most pleasure or the most joy.

Changing or improving conscious thought, changing or improving behavior or changing one's way of thinking or changing point of view is unnatural and is also not very easy. It really requires a lot of effort, a lot of energy and a lot of time for a lot of nothing.

Positive thought really is, therefore, one of those good sounding lies or good sounding half-truths or false science. Positive thinking bears all as it pretends to dress and nurture.

But once the conscious living being or mankind through conscious effort really acknowledges, realizes or really understands or really recollects the real or the actual reason for true, complete or real joy or real and lasting happiness, then true, complete or real joy or real and lasting happiness simply becomes an unnatural but automatic process as conscious thought once became an unnatural but automatic process as the very simple magnetic field or the very simple neutral shield always was an automatic process of physical existence or of physical reality.

In short, once the unnatural but automatic process of true, complete or real joy or real and lasting happiness kicks in, new energy is felt both in body and in mind. At first that new energy will feel like an ecstasy or a natural super high.

In fact, it is really or actually the first enlightenment! It is a true, a complete or a real joy or a real and lasting happiness that keeps coming, even without further effort from the conscious living being, leading to a second and final enlightenment or final acknowledgement.

The very simple anticipation of the next enlightenment really or actually is what makes possible the new found joy or real and lasting happiness.

Ironically, when true, complete or real joy or real and lasting happiness begins automatically as once did conscious thought, the conscious living being really does not understand where it really comes from and so the conscious living being begins to fight real and lasting happiness, as the conscious living being once fought against conscious thought, by feeling odd or by feeling strange, and thus begins to reject that new and automatic and real joy or real and lasting happiness.

As it has been seen and as it has been proven, the universe, Time or simply all of Existence or even physical creation is a product of two simple but infinite magnets in the very simple form of two infinite cubes: The Space Cube and the Matter Cube or call the two infinite magnetic cubes Reality and Non-reality.

By using the Periodic Table of the Elements, which is really a very simple diagram of a bar magnet or a simple diagram of an electric motor or a generator, the above statement, that simply states that all of Existence or that physical creation is really a simple product of two simple but infinite magnets, has simply and thus been proven.

And also, by using the simple and the divisible atom, which really is also a very simple microscopic diagram of a bar magnet or an electric motor or a generator, the first statement of this simple conclusion or this simple continuation has thus simply been proven.

In other words, the true question as to the physical origin of the expanding universe or physical creation has really been simply answered!

And in conclusion or in continuation, the most important thing in all of creation or in all of Existence, seeing or not seeing, once again, simply is or actually is but conscious thought!

Conscious thought only exist in conscious living beings. Conscious thought really is not a natural or is not an automatic process as is all of Existence or as is reality.

Even all of Existence or all of creation as well as non-existence really require something. Conscious thought is not a natural process and conscious thought really requires conscious effort from the conscious living being.

Even though knowledge simply exists or simply "the fact simply and already but is," the conscious living being does not know or does not learn or does not really recollect or does not really remember and thus simply cannot complete Existence or physical creation until the conscious living being makes a conscious effort to simply think and to simply ask that unnatural but true or real question which will simply bring the conscious living being complete or perfect knowledge or acknowledgement or enlightenment!

And the real or the actual purpose of complete or perfect knowledge is simply to let the conscious living being know or really acknowledge that the conscious living being but is and that the conscious living being can really do more with what the conscious living being could really be or could really recognize or could really recollect or could really remember!

Existence as knowledge simply is or simply exists; but ironically, Existence or knowledge knows not. However, just because all of Existence or all of knowledge really knows but not, really does not make Existence or knowledge dumb or useless.

All of Existence or all of knowledge acts or reacts through a simple magnetic or through an electronic impulse or acts or reacts through an actual magnetic instinct.

But knowledge does not know that knowledge really is or that knowledge simply exists. For example, the simple apple tree grows knowing not; but the tree's natural instincts, through knowledge already gotten or already present, is to produce a seed or to simply multiply itself.

The apple of that apple tree ripens or temporarily becomes almost complete not knowing that it does, so that the seed in the almost complete apple can be carried away, but not by the wind or not really by rolling on the ground even though the apple is almost round, but really carried away by living beings.

That is the very simple or the actual reason for the sweetness or the sugar in the almost round apple!

The simple apple in the belly of the living beast or in the hand of the living being is simply or is really a continuation, and not the conclusion, of something that already was.

And even though the apple tree which is really but simple knowledge was really before the seed, the purpose of the apple tree was always the seed, an extension or an expansion of physical knowledge that really already was or was already present!

Another very interesting thing about the knowledgeable apple seed is that the apple seed not only carries all of the physical information of the apple tree, but also the apple seed carries all the knowledge or all the information about the environment or the physical surroundings where the apple tree was or is.

The knowledgeable apple seed no longer has to experiment with the environment or with natural surroundings since the knowledgeable seed already has that physical information.

Instead, the seed as a new or as a second tree will gather more or new knowledge for the next apple or the third tree not yet present.

Whether the new or the second apple from the new or the second tree is sweeter or is more sour than the very first or the previous apple, simply depends, of course, on the physical reaction or the interaction that the living beast or the living being gave to the very first or to the previous apple.

The knowledgeable apple is often colored red to attract the eye of the living beast or the living being with hands, and the redder or the more reflective or shiner the apple, the better for the apple tree and thus the better for the knowledgeable apple seed.

If the natural or the physical surroundings are all red, then the knowledgeable apple with time simply changes to another but more attractive or more attracting color.

By the way, animal seeds or animal cells as do plants seeds have the very same ability or instincts to carry knowledge of previous generations.

But because of the animal's irrational or emotional behavior, or even point of view or new physical location, that first or that previous knowledge or instincts is rarely re-used and usually is forgotten, rejected or even lost.

A very good or an excellent example of lost knowledge or lost instinct is the domestic animal. Left alone in an unfamiliar environment the domestic animal will not survive.

Even house plants lose the ability to survive because of the lack of information or the lack of knowledge of the new environment or new surroundings.

Once again, conscious living beings are really the most important thing in all of Existence, and that is what Existence really reflects or that is what all of Existence really wants or really requires the conscious living being to know.

But not only that!

All of Existence also requires, also wants, or also needs, as the apple tree, the conscious living being to simply act or to react to knowledge, so that all of Existence as well as the living conscious being can be further completed or perfected through conscious thought.

In a way, conscious living beings are actually or are really inside a very large mechanical or infinite magnetic brain or inside a very large electronic brain; and conscious living beings are really the thinking brain cells or living knowledgeable seeds or even living knowledge processors.

Ironically, Existence will not be complete or will not be perfect until all the needs of the conscious living being are simply but met through conscious thought. And that is the actual reality!

The only true or the only complete or the only perfect or real beginning in Time and in all of Existence or even creation is the simple or the actual beginning of conscious thought.

The only limits that the human race, the conscious living being or that mankind really has are the very same limits that the human race, the conscious living being or that mankind really gives or really sets upon himself.

So, therefore, let's get glad and joyfully continue to think and rethink!

Thinking or conscious thought will one day keep the universes or creation as well as mankind from collapsing. Conscious thought will be the conscious shield that will keep the simple universes or creation from collapsing or from imploding into a giant black hole or into a giant grave.

Conscious thought will be the conscious shield or the conscious energy that will one day really keep the simple human conscious mind or the human brain from physically collapsing or from physically imploding also into a hole or into a grave, thus saving the physical universes or saving creation and thus completing or really perfecting physical Existence or physical creation!

The final reality of the matter is that as long as there is conscious thought in the physical universe or physical universes, the physical universes or physical creation will not collapse or will not implode or will not even turn on itself.

The reality of the matter is that there was only one physical creation or only one physical beginning, actually a physical continuation of what already was.

As long as there is conscious thought, only one physical beginning or only one physical continuation or only one physical creation will be so until infinity!

Further Notes

Just as the atom is the greater part of matter or the element, thus our conscious mind is the greater part of the universe or of existence. And just as the electron transforms the atom's magnetic field into an electron field or ring so can we transform our conscious mind into a greater or higher conscious state by eliminating our dead beliefs or our dead faith which will take us dead to the grave if we do not remove our dead beliefs or our dead faith and that includes blind faith from our conscious mind...

Seek the Truth

Seek always of the truth or seek the truth and when she let herself be found from you, you truly will become taller, bright and will even have more confidence in yourself.

The very curious thing about the truth is that she begins with one and if one truly does the movement to find her, thus she will come to one or she will be given to you by he that made or that presented her for the truth.

A Higher Mental Consciousness

The path of less resistance or without contention not only leads to laziness or to vanity or to nothing but also the path of less resistance or without contention leads to death.

And death ends any possibility to any other possibilities, such as a higher mental consciousness or illumination with real gladness and with real joy, in double abundance all five portions, and also with the power and with the authority of the heavens here on the earth.

The lonely road to a higher or taller consciousness

A short auto bio

(1)

A funny thing happened to me on the lonely, narrow and unbeaten road to rebirth or to life renewed, which really is to a higher or to a taller consciousness or an expansion of mental consciousness.

At first, and this was even before I took that first step on the lonely, narrow and un beaten road or path to a higher or a taller consciousness, I wanted to be known.

I wanted to attract countless friends. I also wanted to be popular and perhaps get free lunches or dinners.

But once I took that first step to a higher or taller consciousness, my so called life became lonely, especially when I turned to my so called friends and family.

They marked me, but not before laughing or making jokes.

Some of them even recommended me to professional help.

And so I discovered that one cannot share one's finds with just anyone.

But the irony was that there was not anyone for being something new or the start of a new life.

Most people like to hear of things that they already know about and share or speak of thing that are common or simple, such the news of the day, the weather or social or family events as long as they do not require too much conscious effort from their part.

And so I entered alone into the lonely, narrow and unbeaten road or path to a higher or taller mental consciousness…

(2)

The reason that I walk alone

After trying to share my life improving findings with the people that I liked and even loved, I decided to keep quiet and walk alone but not lonely or feeling alone or even sad on the unbeaten road or path to a higher o taller mental consciousness.

Interestingly, that when I first started my grandiose search and research, I began to feel gladness and joy and even also I felt abundance or a presence...

At that moment I realized that there was more in the world and more to life and so I began to believe or to have faith. I also realized that I needed to acquire or achieve that higher or taller mental consciousness or intelligence.

At first I did not know that I had entered into a higher or taller mental consciousness, but when I began to think of greater or taller things in the world and to see people as they really are, I knew then that I had entered into a higher or taller mental consciousness or state...

I also began to study and read as many books as I could on the subject of a higher or taller mental consciousness, but the close that I ever got was that after life or after death was the answer or the truth.

But I totally rejected that notion or idea or rumor that death was the final frontier…and that more was beyond the grave or after death…

I had read so many books that I realized that most of them were inter connected to each other and that their only purpose was to make someone rich and famous and also to waste one's limited time…

But the books were not the only things wasting one's limited time. Unfortunately, everything else was, from smart cell phones and computers and televisions and eves so called news to even friends and family members!

However, there is no problem with the above matters if one draws a limit or a line and one sets oneself a purpose or a goal to go beyond one's present mental capability.

And that was what I did. I put a limit to everything that did not take me to my purpose or goal, which was to the next consciousness or further into the unbeaten or untraveled road or path.

I even began to physically exercise and take vitamins and to eat as healthy as I could. I lost a lot of weight, mostly fat around by belly.

My waist size went down from 40 inches to 36 inches. My bad cholesterol went down from 300 to 100. My sugar level in my blood before breakfast was 80!

People asked me if I was in love. I answered that there was nothing wrong with that, even though my only love was with my soon to be life renewed.

And some others called me youth or young and smiled at me.

They even told me that I may get a speeding ticket for running so fast!

To be continued...

(3)

I have arrived to an almost unending point

I have arrived to a point in where my wounds have their wounds, my blows have their blows and my scratches have their scratches.

Even my plagues in the form of friends and families have departed from me but no before marking.

But nothing, I have real faith because of searching and researching and contention because of wanting to overcome and become better or taller as also in a better or taller place.

And I know that I will overcome. There is for me no march back because back there is nothing more and what never was and what never will be.

With me there is only one. He has hungered with me in silence.

As the same as me he has also spend nights without sleep because of lack of even a place to rest my feet and my head.

But when the spoils arrive, the spoils so much will be for him as so much will be for me because the spoils will be one, oneself as also the spoils will be himself because of everyone doing for the spoils oneself.

Now then, without contention or struggle or without friction or even without desolation thus there will never ever be overcoming or transformation or there will never ever be a higher place and much less a spoil which will be one oneself as a higher or taller consciousness.

To be continued...

Other Related Works

The Universe

The Universe or the vacuum of space really is a neutral point or point zero or a starting point and the universe becomes positive or turns into a one or a beginning when matter enters the universe or the vacuum of space from the outside of the universe or the vacuum of space.

The universe becomes negative when the largest of stars implode into giants black holes. It is through the giant black holes that the universe or that the vacuum of space is cleared or cleaned from any matter or stars or planets which are still left in the universe or in the vacuum of space and thus returning the universe or the vacuum of space into a neutral point when all of the giants holes are dispersed or turned off because of lack of energy or lack matter in the universe or in the vacuum of space.

Tags: universe, vacuum of space, matter, energy, light, black hole, positive, negative, neutral, beginning, end, and implosion, planets

The Human Brain

The human brain is an exact copy or a duplicate of existence or even God who is a greater or taller consciousness and he that understands existence or God thus understands the human brain and he that understands the human brain understands existence and he will be able to persuade existence or the brain or even God to do his will.

But that persuasion of existence or the brain or even God has everything to do with mental consciousness and he that knows the power of mental consciousness will make existence or his brain his true reality.

In other words, he will be able to really transform existence or transform matter according to his will. However, that requires a clear conscious mind or a conscious mind without false beliefs or a conscious mind with true knowledge

Tags: The Human Brain, cerebrum, mental consciousness, point of view. Police detective, researcher, writer, educated guess, conscious mind, conscious thought, global economy, Stone Age, nightmare, dream, black holes, after death, Paradise

The Black Hole

The Black Hole is a giant hole or a round rupture or tear or breakage in the fabric or the foreskin which really is the vacuum of space.

But the hole or the tear or the breakage in the vacuum of space moves along the vacuum of space sucking up or removing or destroying all matter, including all the stars and planets and even all forms of life, to clear the universe or the vacuum of space for new matter or for another beginning which will appear as if the very first beginning because there will not be any trace that there was ever a first beginning...

Tags: black hole, vacuum of space, stars, planets, beginning, universe, matter, breakage, foreskin, cosmos...

The Origin of the Universe, the complete verse

The Origin of the Universe, the complete version, is about the true and the only origin of the universe.

The tall title says it all. Those conscious beings that do not know reality or do not know the truth thus they will not be able to live in reality or in the truth nor will they live as reality or as the truth they them very selves and they live as if they do not live.

And in truth, they will die because of the lack of reality or because of the lack of the truth, that if recognizing reality or the truth will give them true life and in true abundance of peace, knowledge, joy and gladness.

Tags: Origin, universe, cosmos, existence, matter, space, black holes, space-time, atom, element, table of elements, periodic table of the elements, rings, Seven

My Inspired Words for Royal Life

My Inspired Words for Royal Life are some essays or verses, some short and others long, which explain not only the importance of knowledge and of acknowledgement in the conscious being but also explain how existence truly functions or the grandiose purpose of her or for her and how to truly achieve royal life or the royal salvation by the act of oneself...

Thus in truth, fortunate is he that has the grandiose opportunity of being able to read **My Inspired Words for Royal Life** because he will be amazed and truly will be a better conscious being or will be taller being in identity...

Tags: royal life, royal peace, royal knowledge, royal salvation, the kingdom of the heavens, immortality, royal faith, new heavens, new earth, servant beloved of God, neutral point, purpose, foresters, creation, procreation, adoption, repentance, 01

THE ORIGIN OF THE UNIVERSE

A Short Version

The origin of the universe, finally the last chapter!

All those conscious living beings living today who truly want to know the true origin of the universe, the true origin of life, the true origin of thought and the true origin of the beginning and the end, this is the eBook or printed book to read!

Now then, everything has an answer and the answer starts with one or as one when one questions to know and thus to have knowledge to knowledge thus one becomes, becomes because the purpose of knowledge is thus one as one is thus knowledge...

Thus, truly blessed is that man that has true knowledge of how existence or the universe truly functions, because that man also knows how he himself thus functions.

And he also will do as existence herself to continue on existing or living on but existing or living on as if forever new and in complete, which is perfect, abundance, all five portions of her.

The Real Origin of the Universe

The Real Origin of the Universe is about the true and thus the only origin of the universe.

The tall title says it all. Those conscious beings that do not know reality or do not know the truth thus they will not be able to live in reality or in the truth nor will they live as reality or as the truth they them very selves and they live as if they do not live.

And in truth, they will die because of the lack of reality or because of the lack of the truth, that if recognizing reality or the truth will give them true life and in true abundance of peace, and real knowledge, joy and gladness.

Who am I really?

(1)

My pen name as a writer is Forester de Santos and I am truly on a very grandiose crusade of rebirth alive or to be born again into royal life or life renewed with complete gladness and with complete joy and also with complete abundance of God, which is a higher mental consciousness, but God as much more than God and as much more than Creator.

Now then, one who truly is on a very grandiose crusade cannot follow another or cannot let himself be surrounded by his beloved ones or his fans because he cannot cross over them or he cannot cross over because of them being in the way or because of them blocking the path which is but which cannot be seen until rebirth or until one is born again.

I do not ask to be followed, not because I will not lead, but because I will not look back but I will look to my right and to my left to see who walks with me.

But those that truly decide to follow me will become as me and as me will truly receive or gather true knowledge because my struggle or contention or my very grandiose crusade of rebirth is true, so true in fact that I have become a much better person because of the true faith which I have come to receive through my search and research for the truth.

And because I have come to have true faith or faith of God, which really is a higher consciousness or an expansion of mental consciousness, thus I use my true faith as a shield to repel or to reject other beliefs or good sounding lies!

Therefore, to rebirth alive or to be born again while still living here on the very earth which will be as in the very heavens through rebirth!

Now then, rebirth or life renewed or even God is really about a higher mental consciousness here on earth!

$$[(((((((0+1)))))))]$$

If you truly enjoyed this simple and humble work, please leave a comment according to your good pleasure and give also a rating but also according to your good pleasure.

Thanks so very much for your time and best of wishes, Forester de Santos.

Thanks for reading The Unifying Theory!

0+1 = peace and knowledge to all mankind!

www.ingramcontent.com/pod-product-compliance
Lightning Source LLC
Chambersburg PA
CBHW021815170526
45157CB00007B/2593